Systems Analysis for Computer-Based Information Systems

West Series in Data Processing and Information Systems

V. THOMAS DOCK, *Consulting Editor*

Clark & Mandell	*A Short Course in PL/I and PL/C*
Dock	*BASIC Programming for Business*
Dock	*Structured COBOL: American National Standard*
Dock	*Structured FORTRAN IV Programming*
Graham	*The Mind Tool: Computers and Their Impact on Society*
Mandell	*Computers and Data Processing: Concepts and Applications*
Mandell	*Principles of Data Processing*
Nolan	*Managing the Data Resource Function*
Wetherbe	*Systems Analysis for Computer-Based Information Systems*
Wetherbe	*Cases in Systems Design*

Systems Analysis for Computer-Based Information Systems

JAMES C. WETHERBE
University of Houston

West Publishing Company
St. Paul • New York • Los Angeles • San Francisco

50 West Kellogg Boulevard
P.O. Box 3526
St. Paul, Minnesota 55165

Printed in the United States of America

Library of Congress Cataloging in Publication Data

Wetherbe, James C
Systems analysis for computer-based information systems.

(West series in data processing and information systems)
Includes bibliographies and index.
1. Management information systems. 2. System analysis. I. Title. II. Series.
T58.6.W47 658.4'03 79-1148

ISBN 0-8299-0228-7

In a field as new as computing and information systems, there are few of us who can say—

"My old man is in the business."

I dedicate this book to my father and colleague—

MEAD BOND WETHERBE, SR.

Contents

Instructor's Preface

The disruptive technology of computers has been absorbed by virtually every organization in modern society. Unfortunately, this absorption has evolved with far too few well-trained systems analysts. Managers are becoming increasingly aware that more than just computer skills are needed to bridge the gap between computer technology and effective and efficient use of computer-based information systems in organizations. Individuals who can combine technical skills with organizational skills must be trained. Such individuals are capable of relating with organizations in a manner that facilitates aligning the capabilities of computer technology with the information needs of organizations.

This text has an accompanying case book, *Cases in Systems Design.* Together, these are designed to equip students with both conceptual skills and technical skills necessary for the development of computer-based information systems. As these skills are developed, students are provided opportunities to apply them in a variety of organizational environments.

The first section of the text exposes students to conceptual foundations for systems development, systems theory, information and organizations, and data processing. These conceptual foundations develop students' understanding and appreciation of the role of systems analysis and information in the management of organizations.

The second section of the text presents a comprehensive treatment of techniques used by systems analysts. Students are introduced to well-defined operational frameworks that simplify both the understanding and the application of systems analysis skills. The orientation of the material presented in this section reflects an emerging trend: systems analysis is shifting from purely "technical" analysis towards "information" analysis. That is, emphasis is being placed on "what" information is needed by the organization rather than only on "how" to process it.

The third section of the text presents advanced and future concepts of systems analysis. Major topics include database management systems and decision-support systems.

Throughout all three sections of the text, I have included and emphasized current technology. For example, such topics as distributive processing, online systems, structured programming, chief programmer teams, top-down development, and database management systems are discussed.

Applied Systems Analysis

A common, perplexing problem when teaching systems analysis is the difficulty of providing a "real-world," or applied, orientation. Systems analysis is an applied discipline. The learning experience is compromised if systems analysis is taught in a vacuum.

To address this issue, I have included three significant design considerations in the text and case book. First, extensive use is made of real-world examples, in order to embellish the theories and concepts introduced in the text. Second, midway through the text, after students have learned to prepare systems specifications, a payroll case is introduced. Thus, students are guided to apply the techniques they have learned. Third, after completing the payroll case, students may select more extensive cases from the case book. The case book includes cases in manufacturing, banking, wholesale distribution, and hospital. Approximately 50 percent of all computer installations are concentrated in these four areas. The case book provides unique material for systems analysis instruction which can be used in either of two ways. Under one approach, each student selects one project to complete in conjunction with the text in a one-semester or one-quarter course. Alternatively, the case book can be used in a subsequent semester or quarter, during which time each student completes several projects using the skills learned in the first semester or quarter. This latter approach facilitates student exposure to information systems in several industries. This exposure allows students to more clearly identify areas of career interest. Industry familiarity also prepares students to more effectively interview for systems analysis positions in different industries.

Use of Text

The material in this text is designed for the student who has previously completed or is concurrently taking a course involving a file-oriented procedural programming language such as COBOL or PL/I. It has been successfully used by both undergraduate and graduate students.

Acknowledgments

In conclusion, I wish to acknowledge the contributions of several individuals. I am especially indebted to V. Thomas Dock, the University of Southern California, who encouraged me to write this book and served as consulting editor. The reviewers whose helpful suggestions are reflected throughout the text are: Marilyn Bohl, IBM; Thomas I. M. Ho, Purdue University; Roger Hayen, the University of Nebraska; Steven Alter, the University of Southern California, Richard Scamell, the University of Houston; M. Bond Wetherbe, Information Systems, U.S. Naval Oceanographic Office; Charles Davis, MIS, Occidental Petroleum Corporation; Hank Lautenbach, Idaho State University; and Charles Paddock and Phillip Judd, the University of Houston.

I acknowledge A. Benton Cocanougher, Dean of the College of Business Administration, University of Houston, for his support of my efforts. Sincere thanks to three patient, tolerant, and forgiving secretaries: Susan Burback, Jan Hadley, and Cheryl Patrick.

Finally, I acknowledge the efforts and support of my wife, Smoky, who is a supportive friend and colleague in my professional efforts.

As I acknowledge the efforts of all who have contributed, I also assume full responsibility for any inadequacies or discrepancies in the text.

Student's Preface

The study of systems analysis is an exciting, worthwhile learning experience. The conceptual and technical skills developed by studying systems analysis are highly beneficial, regardless of the career path one chooses—systems analysis, or otherwise.

The three major components required by any organization are material, energy (including human energy), and information. The complexity with which material and energy are combined to produce goods and services in our industrialized society has reached overwhelming proportions. Modern organizations employ thousands of people in thousands of activities. Management and employees often cannot see "the forest for the trees" when attempting to understand all the processes occurring, internal and external to an organization.

Information is one of the critical resources required to run an organization. This resource, taken for granted only a few years ago, is now recognized as essential to the success of an enterprise. It has been estimated that 50 percent of white collar workers' time is spent in locating and processing information needed for decision making. Yet, virtually all decisions are made without adequate information.

The systems analyst equipped with both conceptual skills to understand organizations and their information needs and technical skills to develop information systems to assist organizations has the potential to make a significant contribution to society.

Organization of Text

Chapter 1 provides an overview of systems analysis. Chapters 2 through 4 establish conceptual frameworks for systems, information and organizations, and data processing relevant to systems analysis. The conceptual frameworks developed in the first section of the text prepare the student to develop and apply technical skills of systems analysis. Chapters 5 through 9 are a comprehensive presentation of systems analysis; systems design; systems components; systems development, implementation, and evaluation; and systems administration. The final three chapters present advanced technologies and considerations of systems analysis. Chapter 10 discusses database management systems, Chapter 11 discusses decision-support systems, and Chapter 12 discusses the future of systems analysis.

Systems Analysis for Computer-Based Information Systems

SECTION I Conceptual Foundations

1 Overview of Systems Development

In today's industrialized society, organizations must cope with increasing organizational size, complexity, and uncertainty. The continued success of organizations has become increasingly dependent upon the availability of appropriate information for decision making. Consequently, there has been an increasing need for systems analysts who can design effective and efficient computer-based systems to provide information that is consistent with organizational needs and resource constraints. Computer specialists have consistently represented one of the fastest-growing career areas.

Traditionally, there has been an unfortunate communications gap between the people who use information provided by computer-based information systems and those who design and program them. This gap is generally attributed to the fact that computer specialists often have a good understanding of computer technology but are limited in their understanding of information concepts and how they are applied in organizations. Conversely, the people for whom information systems are designed usually lack computer expertise and may even fear computer technology.

An effective systems analyst can bridge the gap between the computer specialist and the user. However, to do so, his or her repertoire of capabilities should include conceptual skills pertaining to organizations, decision making, human behavior, and information systems. The systems analyst must further possess technical skills in systems analysis and skills in oral and written communications.

The material covered in this text is designed to provide the student with comprehensive preparation in both conceptual and technical skills essential for effective systems analysis. The first part of the text stresses conceptual skills, and the second part stresses technical skills. Approaching information systems analysis in this fashion should allow the student to better understand the "what" and "why" of information systems. The student will then be prepared to learn "how" to design information systems. Systems analysts with such skills know not only how to design an information system correctly but also how to recognize what information system is appropriate.

DEFINITION OF SYSTEMS ANALYSIS

The literal definition of systems analysis is inappropriate when defining what is generally considered the essence of systems analysis. The meaning of the word *system* is much too broad and, conversely, the meaning of the word *analysis* is much too narrow to accurately define systems analysis in the applied sense.

The word *system* can pertain to any system—biological, transportation, stereo, or computer. However, in the context of information systems analysis, the word *system* pertains to computer systems that process information.

The word *analysis* refers to the process of separating the parts of a system to facilitate examining their nature, functions, and relationships. Such analysis contributes to a better understanding of a system. Analysis is certainly a key component of information systems analysis, but it is not the only one. Systems analysis also involves the design, implementation, and evaluation of information systems.

Such variations between the literal and applied use of terminology are not uncommon. For example, the word *manufacture* generally connotes machine-oriented processes of assembling goods or articles. However, the original and pure definition of *manufacture* denotes the making of goods or articles by hand.

For the purpose of this text, systems analysis is defined as the process of analyzing, designing, implementing, and evaluating systems to provide information to support the operations and decision-making processes of an organization. These systems usually imply, but are not necessarily restricted to, computer-based technology. This definition is consistent with the functions performed by systems analysts in organizations.

Role of the Systems Analyst

The systems analyst plays a critical role in an organization. Information and the systems which provide it are among an organization's most valuable resources. The sheer size of most organizations has made it increasingly difficult to keep track of everything occurring in an enterprise. It is a quite common phenomenon for the "left hand" not to know what the "right hand" is doing in an organization. For example, the same hotel room may be committed to two different customers because two different hotel agents are unaware of what the other is doing. Two salespersons may unknowingly commit to deliver the same inventory to two different customers.

The increasing complexity encountered by most organizations can be an even more imposing problem than organizational size. Consider a large piece of industrial equipment assembled from several hundred components. If the information processing pertaining to each component is insufficient to keep one or more parts in stock, production can be stopped for hours or even days. This can seriously impact the profitability of a company.

These dilemmas have greatly increased the difficulty and uncertainty of managing organizations. Information systems are designed to insure the coordination and control of increasingly diverse organizational functions. The systems analyst plays a key role in accomplishing this objective.

THE SYSTEMS DEVELOPMENT CYCLE

All organizations have several types of information systems. At any point in time, varying degrees of these information systems may be computer-based. For example, an organization's payroll information system may be partially computer-based but its accounts receivable information system may be processed manually. As time passes, organizational processes change. The organization's information systems must change in response to the changing needs of the organization. When this is to occur, management usually initiates some form of a systems development cycle to address the problem.

A conceptualization of the basic components of the systems development cycle is given in Figure 1-1.[1] The steps in the cycle are neither discrete nor one-time processes. Rather, the systems development cycle is iterative and evolutionary. The steps in the cycle, however, retain a basic sequential flow from the point of origin of the development cycle— *"new problems and/or opportunities arise."* At any step in the cycle, previously unidentified problems and/or opportunities may be discovered. In such cases, it is important to insure proper integration of a solution to a new problem and/or a new opportunity. Making even a minor change to a system without proper consideration of what has been estab-

1. The components of the systems development cycle are described in varying degrees of detail and categorization by different writers. However, there is general agreement on the flow of the development steps.

lished previously can cause unanticipated and undesirable rippling effects. These effects can seriously damage the operation of the system. Therefore, as Figure 1-1 shows, all steps of the systems development cycle allow return to the point of origin.

New Problems and/or Opportunities Arise

As depicted in Figure 1-1, the impetus for initiating a systems development cycle is the identification of new problems and/or opportunities. For example, a wholesale distributor receives customer complaints concerning late deliveries of orders. These late deliveries result in lost orders and loss of customer goodwill. Action is necessary.

An overview of an order processing system for a wholesale distributor is given in Figure 1-2. The delays in delivery of orders may be caused by several factors (e.g., inefficiencies in processing customers' orders, inventory stock-outs, programming errors.)[2] To clearly understand the exact nature of the problem requires a thorough analysis and understanding of the existing system.

Analyze Existing System

At this point, management may initiate a project team (or task force) to analyze the existing order processing system. This project team can be headed by an individual who is a systems analyst or by a person who is not one but has systems analysis skills.

Analysis of the existing system consists of the following activities:

- Review work flow
- Define decision making associated with work flow
- Review information flow to support decision making (e.g., transactions and reports)
- Isolate deficiencies in the information system

Deficiencies in information systems can be categorized as *inclusion* and/or *structure deficiencies*. Inclusion deficiencies pertain to *what* information, technology, and personnel are included or

2. The delays in delivery of orders can pertain to problems other than those of the information system (e.g., employee negligence or inadequate warehouse capacity). For the sake of brevity, the example used in this introductory chapter assumes the problems are primarily related to the information system.

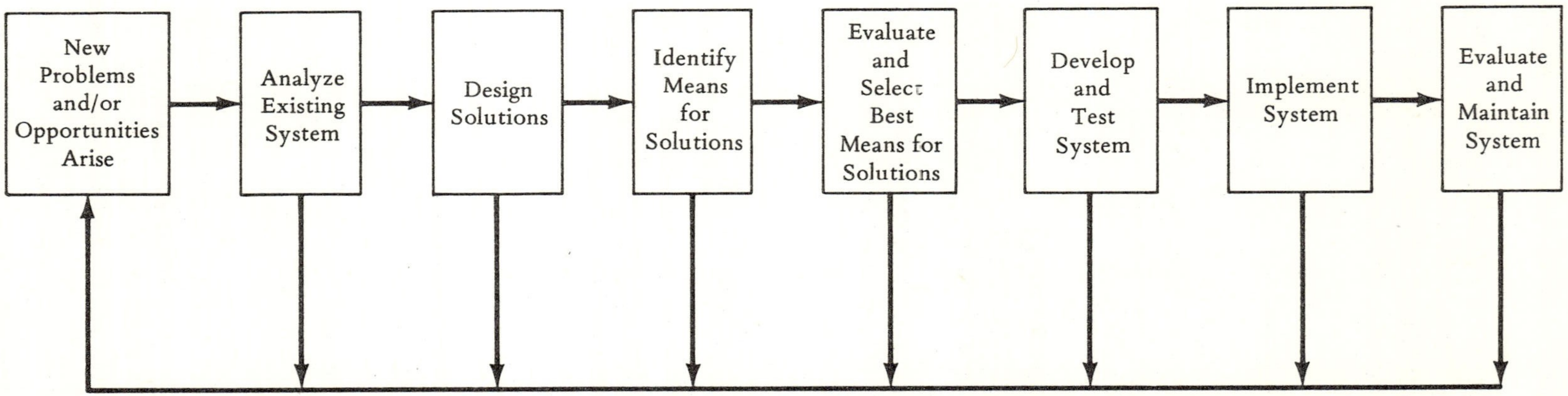

Figure 1-1. *Overview of the Systems Development Cycle*

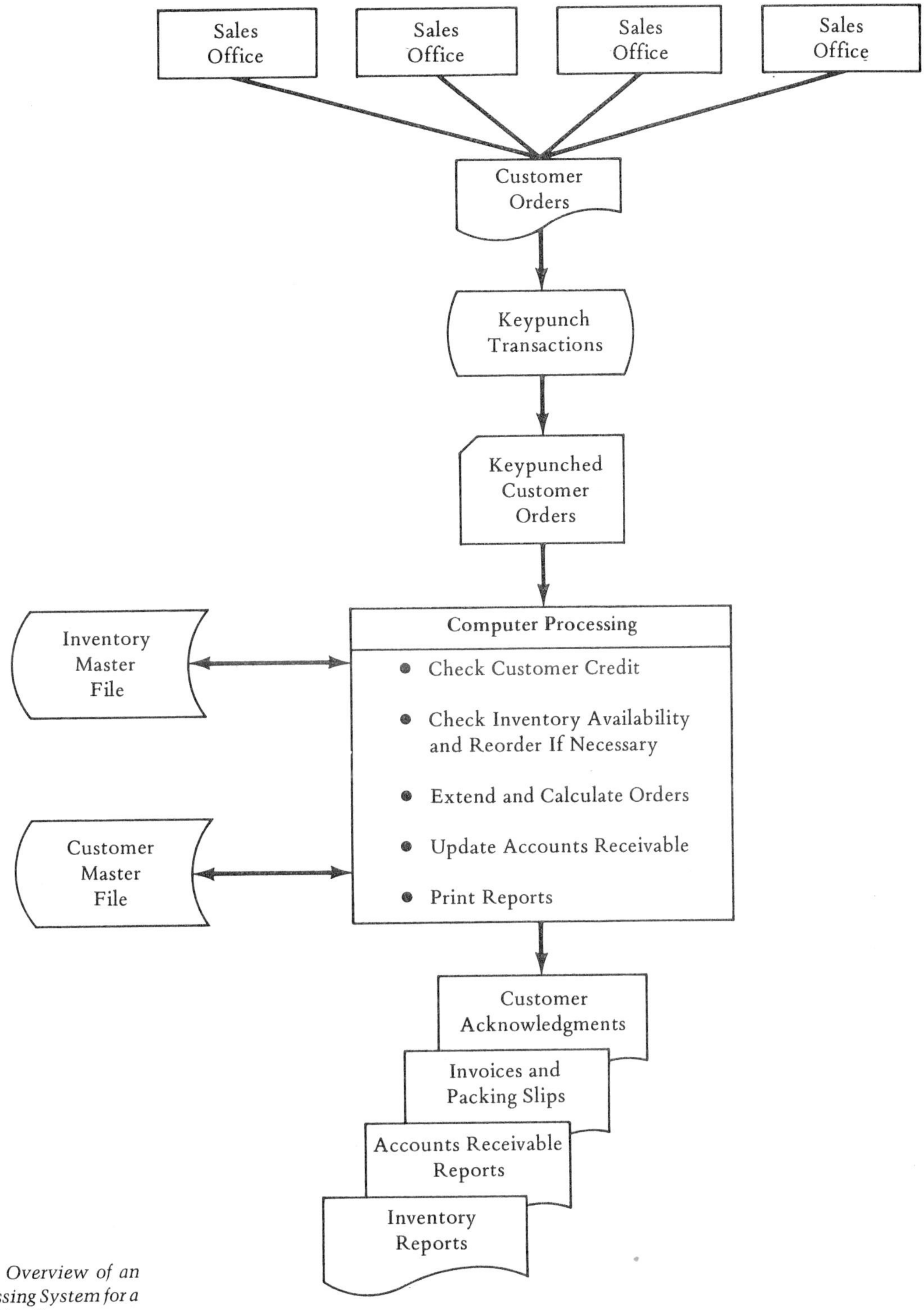

Figure 1-2. *Overview of an Order Processing System for a Wholesale Distributor*

lacking in the system. Structure deficiencies pertain to *how* the information, technology, and personnel are organized and interrelated throughout the system. This concept is illustrated in Table 1-1.

Structure and inclusion deficiencies relate back to a subtle but nevertheless key concept introduced at the beginning of this chapter. This is the concept of both designing an *information system appropriately* and designing the *appropriate information system.* An appropriate information system includes the necessary information, and also the technology and personnel necessary to properly process the information. An information system is designed appropriately when the information, technology, and personnel included in the system are structured and operated correctly. In other words, an appropriate information system is *effective* (i.e., produces the desired information). If an information system is structured appropriately, it is *efficient* (i.e., produces information with a minimum of effort, expense, or waste).

Design Solutions

Once the deficiencies in the information system have been determined, solutions to those deficiencies can be designed. The end result of an information system is information necessary for decision making. Therefore, solutions to information problems can be defined in terms of the inclusion or structure of information.

In the case of the wholesale distributor, it may be determined that historical information about customers' orders would facilitate inventory forecasting. This solution would necessitate the inclusion of new information in the system. A structure-oriented solution could consist of providing a summarized average-demand-by-month report. Since this information is available, all that is required is a restructuring of the existing information into a new form.

In defining solutions, it is important that the solutions relate to improving work flow and/or the decisions being made. Any temptation to include additional information that is not needed, should be avoided. Otherwise, the expense of the information system may be increased needlessly, and managers may be overloaded with voluminous reports that are irrelevant to their decision making. For example, historical information about customer orders that is more than three years old may not be useful for forecasting purposes. To include such information in an information system may increase processing costs and produce useless reports.

Table 1-1. *Information System Deficiencies*

		Inclusion Deficiency	*Structure Deficiency*
INFORMATION	Definition	Appropriate information is lacking, and/or inappropriate information is included in the system.	The manner in which information is collected, stored, and/or reported in the system is difficult or inconvenient to utilize.
	Example	Information that is required for approving a loan application is missing, and/or information that is not required for approving the application is included.	Detail data are not properly summarized.
TECHNOLOGY	Definition	Appropriate technology is lacking, and/or inappropriate technology is included in the system.	The hardware and software included in the system are not organized to most effectively and efficiently process the available information.
	Example	Batch processing is used when an online system is required.	Computer programs are inefficiently written or contain programming errors.
PERSONNEL	Definition	Appropriate personnel are lacking, and/or inappropriate personnel are included in the system.	The personnel associated with the system are poorly organized and/or managed.
	Example	Personnel do not have the aptitude, training, or motivation to work with the system or make decisions from information that it generates.	Personnel responsibilities are not clearly defined, or workload is not evenly distributed among personnel.

Identify Means for Solutions

Definition of the desired solutions establishes the criteria for identifying alternative means for solution achievement. That is, the previous step defines "what" is desired. The step defines "how" to do it. Viable technologies and personnel are identified that, if included in the system, can be structured to support the solution defined in the previous step. Generally, several alternatives offering varying degrees of solution achievement are available. For example, the wholesale distributor currently posts customer orders to the inventory file on a weekly basis. Transactions are delivered to keypunch personnel on a daily basis and are keypunched during the week. The inventory file is updated, and inventory reports are printed and distributed, once a week. The problem with this system is that inventory information is up to a week old. Consequently, sales personnel often overcommit existing inventory. Also, the need to reorder out of stock items is not detectable until the weekly processing has been completed.

One means for improving the currency of inventory reporting is to restructure the existing system to expedite the current processing procedure. Within this alternative, varying degrees of expediting are possible. Transaction posting and report generating could be done twice a week or even daily. Processing twice a week would require an additional computer run. Daily processing would require daily computer runs and would also very likely require either overtime or additional personnel in keypunch.

Inclusion of a new technology is another possibility. For example, online technology could be used for entry of transactions through computer terminals. A less expensive version of this alternative consists of logging transactions onto a cassette tape attached to each terminal. At the end of the day, transactions could be transmitted from the terminal to the computer for further processing and preparation of daily reports. A more expensive version of this alternative is to have terminals connected directly to a computer system. As each transaction occurs, inventory information is updated immediately. This way, the quantity-on-hand information would always be up-to-date.

It is important to identify all viable means of resolving the problem under consideration. In the technologically dynamic field of computers, particular care must be taken not to simply resort to previously used approaches that may already have become obsolete through the development of more effective and/or more efficient technologies.

Evaluate and Select Best Means to Solutions

Evaluation and selection consists of carefully considering the possible approaches to problem resolution in a cost-versus-benefit framework. Sometimes costs can be accurately estimated, however, it is generally difficult to accurately assess the monetary benefits of improving an information system. For example, how much benefit will be derived from improved good will due to expediting customer orders? Consequently, selection of one system over another is often based on subjective evaluations.

Management involvement in the selection decision is particularly important. Management has the best "vantage point" from which to make subjective-type decisions within the organization.

Develop and Test System

At this point, the desired solutions and the means of achieving them have been identified. Specifically, inclusion and structure decisions are established for the information and the technology of the new system. The actual structuring, or development and testing, of the system is now possible. This step consists of installing any additional hardware or software required and coding and testing computer programs. The end product of this step is an operable and, hopefully, reliable system that actualizes the original design objectives.

Implement System

After the new system has been developed and tested, conversion from the old system to the new system is possible. The actual implementation process can be expressed in terms of a continuum that ranges from parallel to discrete.

In a parallel implementation, the old and new systems are concurrently processed until the new system stabilizes. This insures the reliability of the new system prior to abandoning the old one. In a discrete implementation, the old system is terminated as the new one is begun. In many cases, some parts of a new system are implemented in parallel, whereas others are implemented in a discrete manner.

Parallel processing reduces risk when implementing a new system. It does, however, have disadvantages. First, parallel processing is more expensive than discrete implementation. Often, additional personnel (or overtime) are required and/or additional equipment is needed. For example, assume the wholesale distributor (above) is converting from batch to online processing of customer orders. To parallel process during the implementation will require preparing the transaction documents for keypunching (the old system) in addition to entering the transactions through terminals (the new system). Keypunch machines and keypunch personnel and the computer processing and report generating functions of the old system will still be required. If the computer that supports the online processing is to replace a computer in the old system, both the new and the old computers will be required until conversion is completed and parallel processing is terminated.

Even if warranted, parallel processing is occasionally not practical. For example, it is not plausible to parallel process two online airline reservation systems. In such cases, rigorous testing of the new system is critical prior to attempting a discrete implementation.

Evaluate and Maintain System

After a new system has been implemented, it is important to review how effectively and efficiently the solutions to the new problems and/or the new opportunities have been achieved. Evaluation, therefore, consists of assessing the degree of variation between planned and actual systems performance.

This evaluation relates back to the concepts of inclusion and structure. If the availability and the form of the information are consistent with design specifications, then the appropriate information is included and it is structured appropriately. This can be determined by surveying users of the information system to determine if they do, in fact, have the information necessary for decision making. The information included in the system can also be evaluated by assessing decision-making performance. For example, does having information available on average-demand-by-month reduce inventory costs by 10 percent?

If the timeliness and accuracy of the information are consistent with design specifications, then the appropriate technology and personnel are being used and they are structured appropriately. For example, are computer terminal operators consistently able to achieve response times of less than two seconds for queries made to the inventory file?

If the new system seriously fails to achieve the design objectives or presents new problems or opportunities, a new systems development cycle may have to be initiated. If the new system performs satisfactorily, then the system can be maintained at the current operating level until new problems and/or opportunities arise.

SUMMARY

The need for more timely, accurate, and complete information is continually increasing as organizations increase in complexity and size. Accordingly, there is an increasing need for systems analysts who can assist in bridging the gap between computer technology and organization needs. The role of the systems analyst in designing information systems is analogous to the role of the architect in designing a building. He or she must develop an understanding of, and an intimacy with, the organization's problems and work with the organization to design appropriate solutions.

The systems development cycle consists of eight steps. It evolves and recycles with time. The steps in the cycle are:

- New problems and/or opportunities arise
- Analyze existing system
- Design solutions
- Identify means for solutions
- Evaluate and select best means to solutions
- Develop and test system
- Implement system
- Evaluate and maintain system

Throughout the systems development cycle, the inclusion and structuring aspects of information, technology, and personnel must be carefully considered to achieve an effective and efficient information system.

EXERCISES

1. Define systems analysis in the information systems context.
2. Discuss the reasons for the increasing demand for information in organizations and the role of the systems analyst in addressing this demand.
3. Define and discuss the steps of the systems development cycle.
4. Differentiate between inclusion and structure as they pertain to information, technology, and personnel.
5. Visit with several managers, preferably in different organizations, and question them as to the deficiencies in the information available to them. Categorize these deficiencies according to the matrix in Table 1-1.

SELECTED REFERENCES

Ackoff, Russell L. "Management Misinformation Systems," *Management Science,* Dec 1967, pp. 147–156.

Anderson, William S. "Bridging the Systems Expectation Gap," *Data Management,* Nov 1975, pp. 29–32.

Business Week Editorial. "Business Takes a Second Look at Computers," *Business Week,* June 5, 1971, pp. 59–136.

Kanter, Jerome. "Impact of Computers on the Business Organization," *Data Management,* Apr 1972, pp. 20–23.

Keider, Stephen P. "Why Projects Fail," *Datamation,* Dec 1974, pp. 53–55.

Larson, Harry T. "EDP: A 20-Year Rip-Off," *Infosystems,* Nov 1974, pp. 26–30.

Lucas, Henry C., Jr. "A Descriptive Model of Information Systems in the Context of the Organization," *Data Base,* Winter, 1973, pp. 27–39.

McFarland, Warren F. "Management Audit of the EDP Department," *Harvard Business Review,* May–June 1973, pp. 131–142.

McFarland, Warren F., Richard L. Nolan, and David P. Norton, *Information Systems Administration.* New York: Holt, Rinehart and Winston, 1973.

McKinsey & Company, Inc. "Unlocking the Computer's Profit Potential," *The McKinsey Quarterly,* Fall 1968, pp. 17–31.

Nolan, Richard L. "Business Needs a New Breed of EDP Manager," *Harvard Business Review,* Mar–Apr 1976, pp. 123–133.

Nolan, Richard L. "Plight of the EDP Manager," *Harvard Business Review,* May–June 1973, pp. 143–152.

Nugent, Christopher E., and Thomas E. Vollman, "A Framework for the Systems Design Process," *Decision Sciences,* 1972, pp. 84–109.

Sprague, Christopher R. "Barriers to Progress in Information Systems Design: Technological Or Not," *Data-Base,* Winter 1973, pp. 115–120.

2 Concepts of Systems

The word *system* is used widely. It has become fashionable to attach the word *system* to add a contemporary flair when referring to things or processes. People speak of exercise systems, investment systems, delivery systems, and information systems. This chapter introduces general systems concepts with emphasis placed on concepts used later in the text in discussions of analyzing and designing information systems.

DEFINITION OF A SYSTEM

In the most general sense, a system is a collection or arrangement of *entities*, or things, related or connected such that they form a unity or whole. In systems designed and controlled by people, the entities are generally arranged so that they can interact to accomplish one or more objectives. For example, the entities of an educational system consist of students, faculty, administrators, textbooks, buildings, equipment, and so forth. The entities of the system interact (e.g., faculty teach students, students read texts, administrators schedule classroom facilities), and ultimately the primary objective of the system—learning—is accomplished.

Usually, some entities are retained in (internal to) the system and other entitites are transient to it. Transient entities are inputted to the system and subsequently outputted from the system—generally, after some transformation process has occurred. During the time that transient entities are in the system, they are part of the system. However, prior to being inputted or after being outputted, transient entities are external to the system. Anything external to a system is referred to as the *environment* of the system.

In an educational system, the entities retained in the system include facility, administrators, textbooks, buildings, equipment, and the like.

Students are transient entities, inputted to and outputted from the system. The primary transformation processes consist of transforming students with one level of knowledge to students with a higher level of knowledge.

General Model of a System

It is helpful to view systems in terms of models. A basic model of a system consisting of inputs, outputs, and the system itself is shown in Figure 2-1*a*. As previously defined, a system is composed of entities that interact to achieve a transformation process. An expanded version of the basic model depicting system entities is given in Figure 2-1*b*. The expanded model of a system can be applied to an educational system as shown in Figure 2-1*c*.

Classification of Systems

Systems can be viewed or classified in several ways. For the purposes of this text, the most useful classification scheme consists of the categories of closed/stable/mechanistic systems and open/adaptive/organic systems. Few systems are exclusively categorized

Figure 2-1. *(a) Simple Model of a System. (b) Expanded Version of a System. (c) Model of an Educational System*

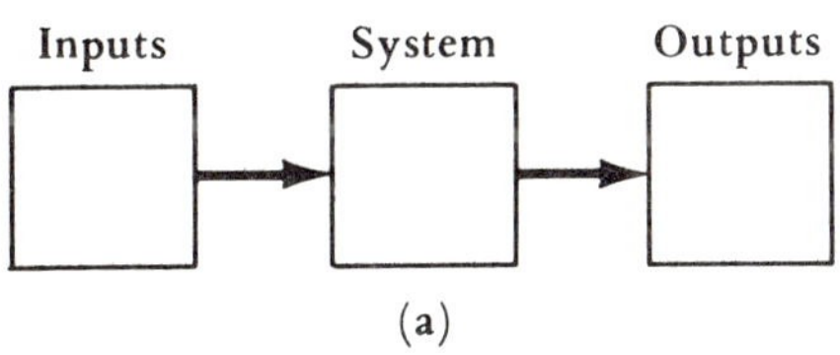

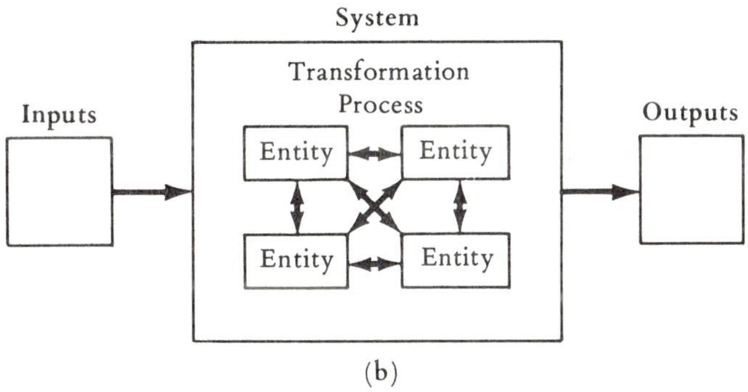

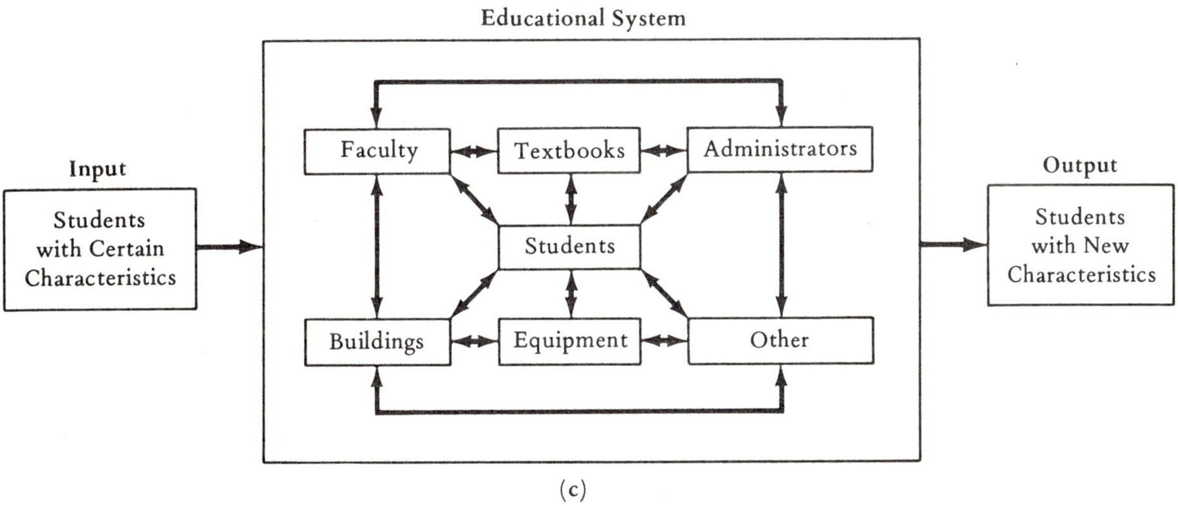

one way or the other. Rather, any given system can be defined as existing on a continuum between the two categories. This concept as applicable to three systems (A, B, and C) is illustrated below.

System A *System B* *System C*

Closed/Stable/Mechanistic ←——◆——————◆——————◆——→ Open/Adaptive/Organic

Closed/Stable/Mechanistic Systems

A closed/stable/mechanistic system tends to be self-contained. It seldom if ever interacts with its environment to receive input or generate output. Therefore, the entities of primary concern in a closed/stable/mechanistic system are those retained in the system. The interactions of the entities are stable and predictable. Consequently, the operation of the system tends to be highly structured and routine. A model of a closed/stable/mechanistic system is given in Figure 2-2.

An example of a closed/stable/mechanistic system is a terrarium (a glass container enclosing a garden of small plants). The primary entities in this system include plants, moisture, soil, oxygen, and carbon dioxide. These entities interact with each other in a very stable and predictable fashion.

No known system can continue to operate for prolonged periods of time without interacting with its environment. Systems are subject to deterioration or slow decay. They eventually need to input new entities of material, energy, or information in order to survive. In the case of the terrarium, water occasionally needs to be added so that the system can continue operation. Another example

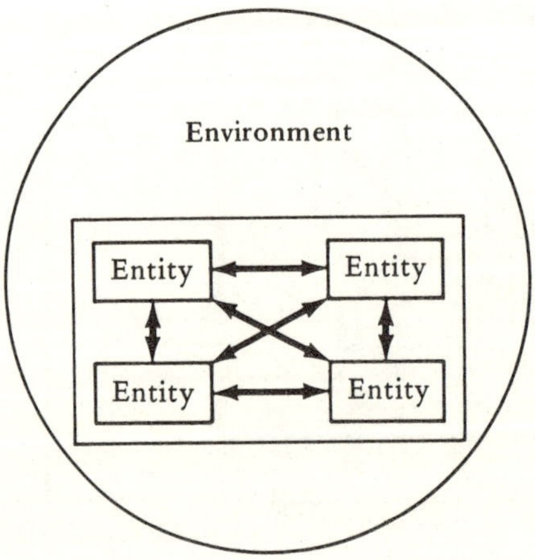

Figure 2-2. *A Closed/Stable/Mechanistic System. (No inputs or outputs.)*

of a closed/stable/mechanistic system is an electric motor powered by a battery. This system can continue to operate without interacting with its environment until the battery runs down or the motor needs repair.

A closed/stable/mechanistic system does not necessarily have to interact with its environment to continue to exist. However, in the long run, it will have to interact to continue operation.

Open/Adaptive/Organic Systems

An open/adaptive/organic system continually interacts with its environment for replenishment of material, energy, and information. Therefore, in an open/adaptive/organic system, entities internal and external to the system are of concern. The interactions of these entities are probabilistic and changing in nature and are consequently less predictable than those in a closed/stable/mechanistic system. The operation of an open/adaptive/organic system tends to be less structured and routine than that of a closed/stable/mechanistic system. An excellent example of an open/adaptive/organic system is a business organization.

A distinctive feature of an open/adaptive/organic system is that it can adapt to changing internal and environmental conditions. It can change the entities retained in or transient to the system. It can also change the way that entities interact. An open/adaptive/organic system is self-organizing, and it can change its organization in response to changing conditions. For example, a business organization can change its personnel, equipment, or products; it can also change the way in which they are organized and what they do.

Effect of Inputs and Outputs

Though all systems that continue to operate have to interact with their environment, the frequency of this interaction is a variable that plays a key role in defining the position of a system on the continuum between closed/stable/mechanistic and open/adaptive/organic. The other key variable associated with interaction is the characteristics of the inputs and outputs to the system. If the inputs and outputs are known, defined, and predictable, even though the system is not closed, it is still relatively stable and mechanistic. Alternatively, if some of the inputs and outputs are unknown, undefined and unpredictable, then the system has to be

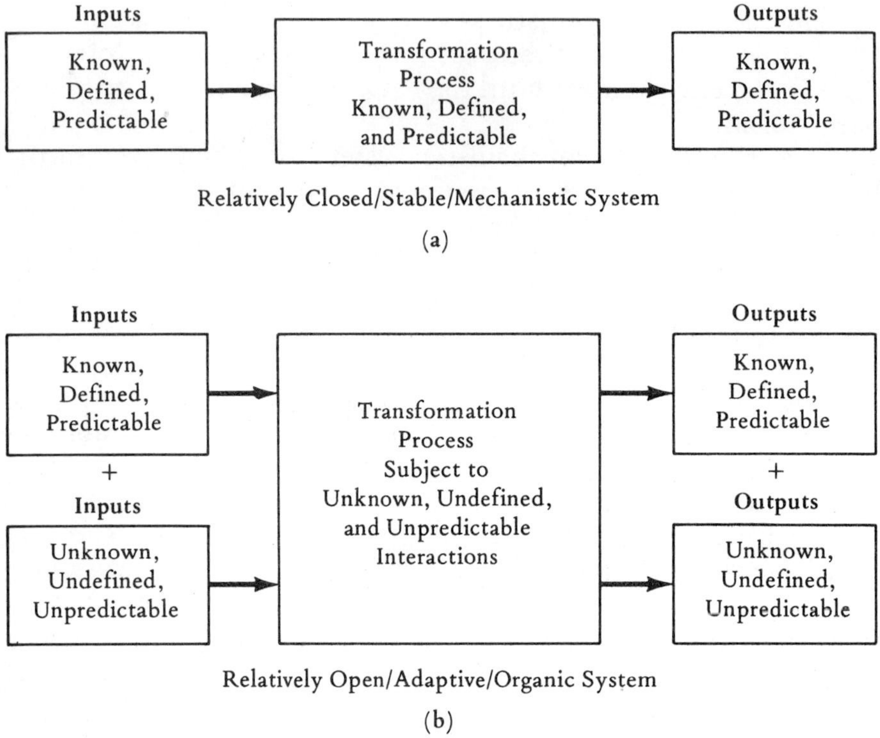

Figure 2-3. *Illustrations of (a) Closed/Stable/Mechanistic and (b) Open/Adaptive/Organic Systems*

adaptive and organic to cope with these unknown variables. Illustrations of relatively closed/stable/mechanistic and open/adaptive/organic systems are given in Figure 2-3.

An example of a relatively closed/stable/mechanistic system is an internal combustion engine. The inputs are gasoline and air, and the outputs are power and exhaust. The input, output, and transformation processes are stable and mechanistic.

An open/adaptive/organic system can be illustrated by the educational system discussed earlier. Students, which are the primary input to the system, are variable and often unpredictable. The best way to achieve the transformation process (learning) is not highly defined, and, of course, the characteristics of the outputs are variable.

System Control

To insure the proper operation of a system, some form of control is required. To accomplish this, a feedback process must be added to the basic system model. A system without feedback for regulatory purposes cannot intelligently adapt to change internal or external

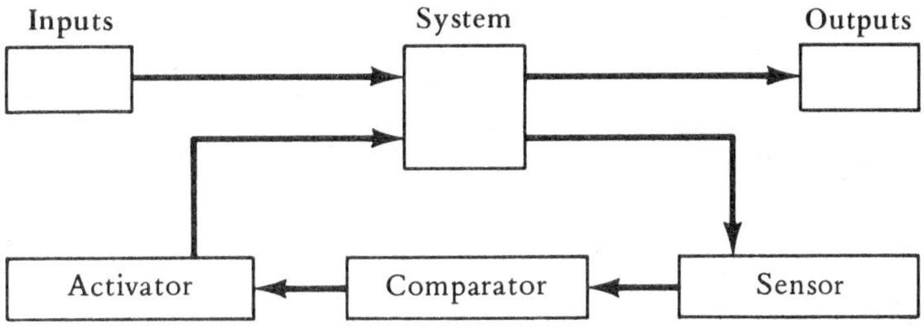

Figure 2-4. *The Addition of Feedback to a Basic System Model*

entities to the system and is subject to entropy or slow decay. The addition of feedback to the basic system model (Figure 2-1*a*) is shown in Figure 2-4.

This model illustrates that controlled characteristics or conditions are sensed or measured by a sensor. A comparator is utilized to compare actual conditions to desired conditions. Finally, an activator invokes action to bring about change (if necessary) in the operation of the system.

An example of a system with feedback is a thermostat-controlled heating system. Once the desired temperature has been set, the system continually senses the room temperature and compares it to the thermostat setting. When necessary, the processes to either turn on or turn off the heating system are activated.

Organizations require a multiplicity of control systems to monitor performance. For example, operating costs are compared to budgets. Sales and marketing research provide feedback on how well products are being received. Employees are given feedback through annual performance reviews.

A system operating without feedback is analogous to a person hitting golf balls in the dark. Without visual feedback on performance, whether or not adjustments are necessary for improvement cannot be determined.

SUBSYSTEMS

The preceding discussion of systems considered only one system at a time. In actuality, the world is full of systems. Every system except the suprasystem is part of a larger system. The *suprasystem* contains *all* systems. This suprasystem is yet to be defined, but it at least contains the universe and whatever else there might be.

Since every system is part of a larger system, it is also a *subsystem* of a larger system. For example, a sales department is a subsystem of an organization. The organization also contains other subsystems such as production, engineering, and accounting departments. The organization can be viewed as a subsystem of a particular industry that includes other organizations in the same business. The industry is a subsystem of all industries. This process of defining larger systems can be continued until the suprasystem is reached.

In order to study or design a system, it is important to define what system is under consideration. Though the primary idea of the systems concept is to work with the whole system, this is often not practical. It is often helpful to simplify the study or design task by factoring the system into subsystems. The interactions among subsystems can then be defined in the same way that the interactions among entities are defined. This concept is illustrated in Figure 2-5.

It can now be pointed out that an entity is actually a subsystem that is not further simplified. For example, one of the subsystems of a sales department is personnel. Personnel also have subsystems such as circulatory and respiratory subsystems, which have further subsystems, all the way down to molccular subsystems. However, for most analysis, considering a particular subsystem as an entity (without additional factoring) is quite adequate.

Interface

The interaction among subsystems consists of the exchange of inputs and outputs of material, energy, and information. With knowledge of what inputs go into and what outputs result from the operations of a particular subsystem, the subsystem can be *interfaced,* or connected to other subsystems which either generate inputs needed by the subsystem or consume outputs generated by the subsystem. This concept is illustrated in Figure 2-6.

In order for the output of one subsystem to be acceptable to another subsystem, the interface between the two subsystems must be operational. That is, the output from one subsystem must adhere to certain standards to insure that it is acceptable as input to the next subsystem. For example, the interface between electrical power and an electrical appliance (i.e., the wall socket and the plug) is standardized so that the two subsystems can be connected.

The more rigidly standards are adhered to, the more conveniently subsystems can be interfaced. In the case of the electrical appliance, if variously shaped plugs and wall sockets are used, achieving interface requires intervention in the form of changing plugs or sockets.

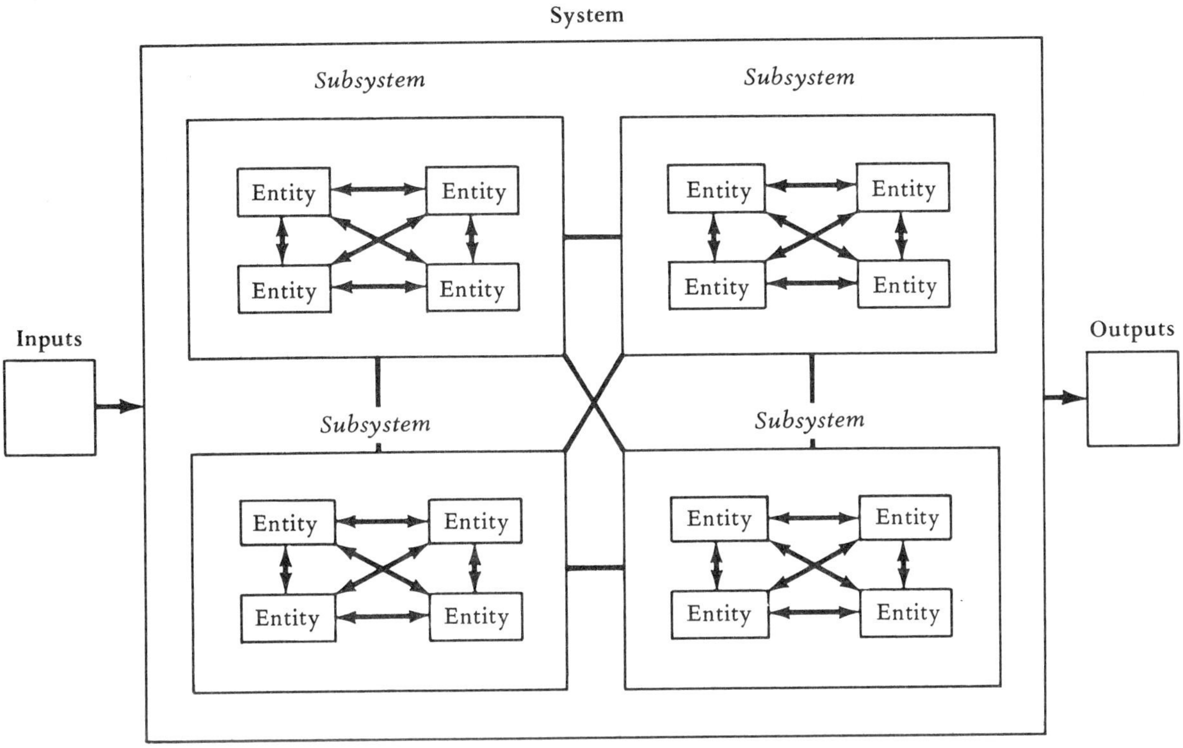

Figure 2-5. *System Composed of Subsystems*

It is usually difficult to maintain rigid standards for interfacing many systems (or subsystems). Consequently, various *adapting techniques* are used to allow more flexibility in interfacing one system with another. The two general categories of adapting interfaces are *translation* and *slack resources.* Translation consists of converting an output of one system into an input form suitable for the next system. For example, a plug adapter can be used to interface an incompatible wall socket with a plug.

A common problem when interfacing systems results from the inability of one system to consistently generate output at a rate that optimizes the performance of the next system. For example, the manufacturing department cannot produce just the right amount of output to accommodate the sales department. Sales

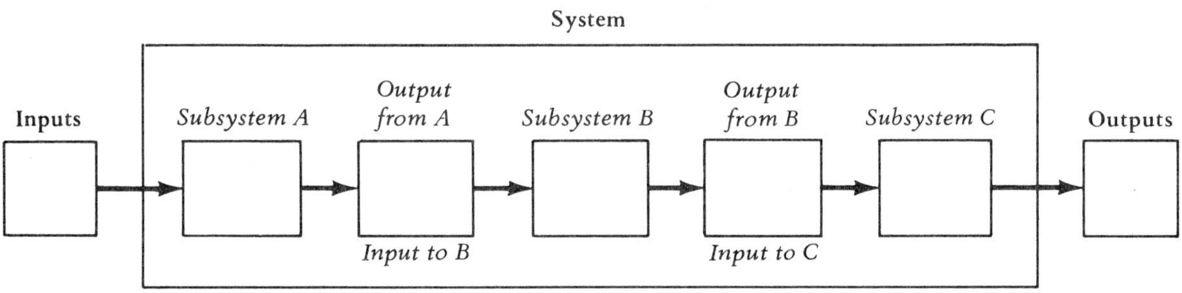

Figure 2-6. *Interfaces between Subsystems*

demands are usually cyclical, and, therefore, slack resources are required to maintain the desired interface between the systems. Slack resources are achieved by maintaining inventories and/or maintaining excess capacity. The use of inventories insures the availability of sufficient output from one system to be used as input to another. Inventories absorb variations in demand. Excess capacity allows a system to increase output to accommodate increased demand from another system.

The more slack resources available, the more expensive (i.e., inventory carrying costs and/or underutilized capacity) the interfacing between systems becomes. However, poor interfacing between systems can result in a system not being able to operate due to the unavailability of inputs. There are costs associated with such losses in system productivity. Therefore, the costs of interfacing must be balanced against the costs of curtailing or stopping a system. For example, the costs of having excess capacity in a factory are offset by the revenues generated from being able to accommodate fluctuations in customer demand.

The interfacing pocesses of translation and slack resources are achieved either by adding them to the existing systems or by creating an additional subsystem to handle the interface. These concepts are illustrated in Figure 2-7.

Figure 2-7. *(a) Interface by Adding a Function to One or Both Systems. (b) Interface by Creating Additional Subsystem*

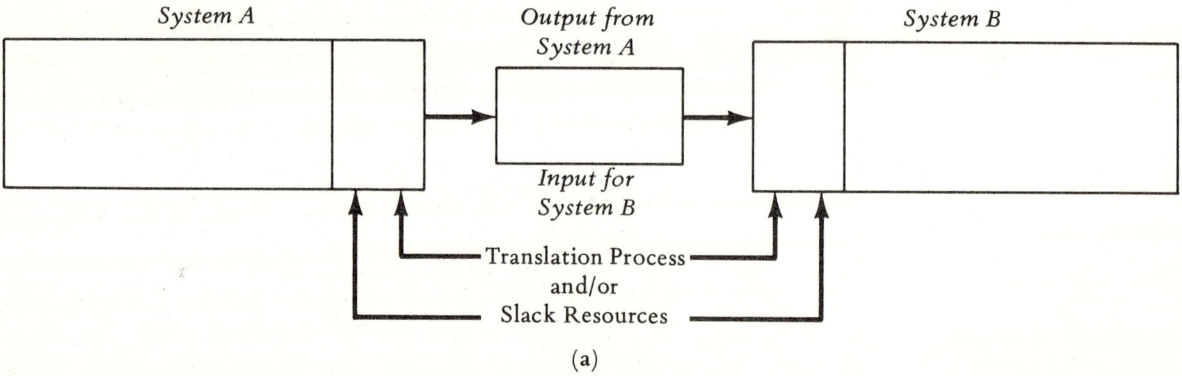

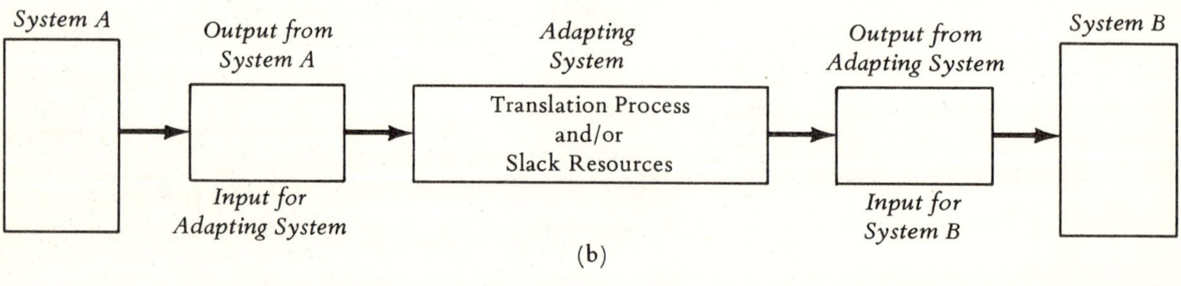

THE SYSTEMS APPROACH

The term *systems approach* emerged in the 1950s as a label for what was becoming a more analytical orientation to management and problem solving. During that era, Kenneth Boulding provided a perspective of the systems approach which is still commonly used today:

> The systems approach is the way of thinking about the job of managing. It provides a framework for visualizing internal and external environmental factors as an integrated whole. It allows recognition of the function of subsystems, as well as the complex suprasystems within which organizations must operate. Systems concepts foster a way of thinking which, on the one hand, helps the manager to recognize the nature of complex problems and thereby to operate within the perceived environment. It is important to recognize the integrated nature of specific systems, including the fact that each system has both inputs and outputs and can be viewed as a self-contained unit. But it is also important to recognize that business systems are a part of larger systems—possibly industrywide, or including several, maybe many, companies and/or industries, or even society as a whole. Further, business systems are in a constant state of change—they are created, operated, revised, and often eliminated.[1]

The systems approach is based upon general systems theory, which is concerned with the development of a systematic theoretical framework for the empirical world. The ultimate (and currently unachievable) goal of systems theory is to define a framework that ties all disciplines together in a meaningful relationship.

Implementation of the systems approach draws upon (1) systems philosophy, (2) systems analysis, and (3) systems management.[2] *Systems philosophy* is involved with thinking about or conceptualizing phenomena in terms of wholes consisting of entities or subsystems with emphasis placed on their interrelationships. *Systems analysis* involves methods or techniques of problem solving or decision making. Systems analysis is further concerned with becoming aware of the problem, identifying the relevant variables, analyzing and synthesizing the various factors, and determining an optimum or at least satisfactory solution or

1. Kenneth Boulding, "General Systems Theory—The Skeleton of Science," *Management Science*, Apr 1956, pp. 197–208.

2. Richard A. Johnson, Fremont E. Kast, and James E. Rosenzweig, *The Theory and Management of Systems*, 1973.

program of action. *Systems management* is concerned with the application of systems theory to managing organizational systems and/or subsystems.

The systems approach can thus concurrently be considered (1) a way of thinking, (2) a method or technique of analysis, and (3) a managerial style. The integration of these concepts is illustrated in Figure 2-8.

The traditional scientific approach to problem solving involves isolating and simplifying a system so that its various elements can be analyzed and improved. This process, a process of reductionism, plays a critical role in problem solving. Once a system has been reduced to its simplest form, problems with individual system elements can be more readily identified and resolved. However, reductionism alone is not adequate when studying many systems of major concern today.

A vast, complex system has many elements with significant interactions that are often not understood. Isolating the elements for analysis may lead the systems analyst to overlook the interactions that occur when the elements are combined into a system. (For example, understanding hydrogen (H) and oxygen (O) does not insure understanding of the combination of H_2O—water.) In particular, a system that involves human behavior (e.g., an organization) involves varying and changing variables that interact with,

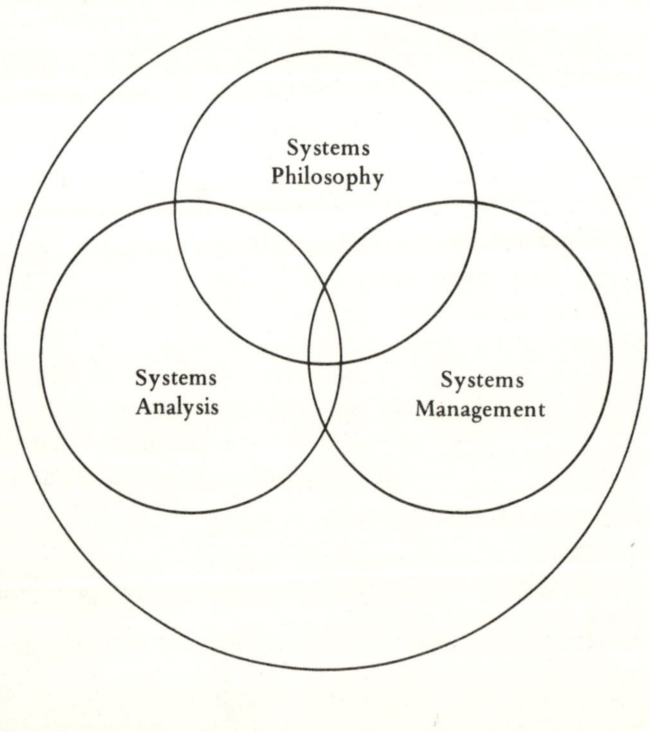

Figure 2-8. *Integration of Systems Concepts within a Systems Approach*

and/or compromise, the system. One of the key realizations contributed by the systems approach is that the interactions among elements in a system can be as or more important than the elements themselves. A system is more than just the sum of its parts and must be considered as such.

DESIGNING SYSTEMS

One of the most useful and structured frameworks for designing systems was developed by Christopher E. Nugent and Thomas E. Vollman.[3] Within their framework, the first step in designing a system involves clearly establishing the criteria and objectives of the system. The second step involves discovering which system or systems to use to satisfy the systems objectives. The search for the desired system constitutes the core of the design process. It involves:

1. Deciding what entities to include in (or exclude from) the system. This is referred to as the *inclusion* process.
2. Deciding how to structure the attributes of the set of chosen entities. This is referred to as the *structuring* process.

The inclusion process logically precedes the structuring process. However, in practice the systems analyst alternates between the inclusion and structuring processes. Inclusion is a highly creative process wherein the systems analyst must look beyond the obvious entities and search for new and significant entities possessing attributes that can better contribute to the performance of the system. For example, a soccer player's (an entity) kicking ability (an attribute) has been found to be most effective for field-goal kicking (an objective) in football (a system).

Exclusion is a necessary part of systems design. It is frequently a difficult and frustrating process—for example, when some relevant entities have to be excluded from a system to constrain complexity. A systems analyst may recognize the potential contribution an entity offers to a system but not be able to deal with a system that includes the entity. For example, inclusion of solar energy in transportation systems is desirable but not yet practical.

3. Christopher E. Nugent and Thomas E. Vollman, "A Framework for the Systems Design Process," *Decision Sciences,* 1972, pp. 84–109.

The structuring process involves organizing the attributes of the entities included in the system. This involves determining the key relationships among the attributes of the chosen entities. The structuring process is important and yields payoffs. However, often, too much effort is directed toward improvements in structure (e.g., queueing theory, mathematical programming, or game theory) when inclusion changes might yield higher payoffs. Our educational process is often considered to be the reason why most problem solvers emphasize structure without sufficiently addressing the question of inclusion. For example, Russell L. Ackoff has noted the tendency of educators to stress exercises, or mechanized problems, rather than total problem solving.[4] Recognizing when to emphasize inclusion and when to emphasize structure is a key to good systems design.

One final element of good systems design is accommodating systems that are dynamic by explicitly and implicitly considering a time dimension. Inclusion of the time dimension denotes that a system is designed to perform over time. A system should bc able to evolve over time and in some cases to accelerate that evolution.

SUMMARY

A system is a related collection of entities generally designed to accomplish multiple objectives. Systems receive inputs that are processed to produce outputs.

Systems can be classified as being either closed/stable/mechanistic or open/adaptive/organic in orientation. Closed/stable/mechanistic systems have more routine and predictable interactions among entities and subsystems internal to the system than do open/adaptive/organic systems. Closed/stable/mechanistic systems either do not interact with their environment or interact in a more predictable fashion than open/adaptive/organic systems.

System control is achieved by some form of feedback process. Feedback is compared to standards or expectations to determine adjustments needed to achieve or maintain system objectives. Systems are composed of subsystems that must be properly interfaced. If necessary, translation processes or slack resources can be introduced.

---

4. Russell L. Ackoff, "Computer Obstructed Education," *Society for Management Information Systems*, 1973, pp. 108–114.

The systems approach is a way of thinking about analyzing and managing systems. It is not concerned only with the parts of a system; rather, it considers the total effect created when the parts function as a whole—the system itself. Designing systems consists of establishing systems objectives, selecting for inclusion those entities possessing attributes that can contribute to the systems objectives, and appropriately structuring the entities included in the system.

EXERCISES

1. Define and discuss the word *system.*
2. In the following list of entities relevant to a manufacturing concern, classify each entity as being primarily internal, transient (i.e., inputs and outputs), or external to the manufacturing system.
   a. Equipment used in a plant
   b. Raw materials
   c. Assembly-line workers
   d. Customers
   e. Manufacturing facility
   f. Management
   g. Competitors
   h. Government regulations
3. Classify the following systems as being primarily closed/stable/mechanistic or primarily open/adaptive/organic.
   a. Clock
   b. Human being
   c. Fraternity
   d. Prison
   e. University
   f. Mathematics
4. Discuss the use of feedback in controlling the performance of systems. What forms of feedback can be used to control the performance of the following systems or system entities?
   a. A student
   b. A football team
   c. An employee
   d. A manager
   e. An organization
5. Discuss the relationships among systems, subsystems, and entities.
6. The interfacing of two subsystems is achieved by the use of standards and/or adapting techniques (i.e., translation or slack resources). Dis-

cuss interfacing techniques as they may be used in the following environments.
   a. Interfacing the manufacture of seasonal goods (e.g., lawn mowers) with consumer demand
   b. Interfacing stereo components
   c. Interfacing high school education with college education
   d. Interfacing the manufacture of automobile replacement parts (e.g., tires and batteries) with automobiles requiring them
   e. Interfacing ambulance services with a volatile demand for ambulance service

7. Discuss the difference between the traditional approach to problem solving (i.e., reductionism) and the systems approach.

8. Determine whether the following problems are inclusion oriented or structure oriented.
   a. A college has an unusually high dropout rate due to admission standards that allow students who are not qualified to enter the college.
   b. A college has an unusually high dropout rate due to poor advising of students. Because students are not guided to take courses in the proper sequence (e.g., calculus is taken before algebra), they frequently become discouraged and leave the college.
   c. A company is losing sales of a new product because there is little demand for the product.
   d. A company is losing sales of a new product because the product is assembled poorly and is therefore of inferior quality.

SELECTED REFERENCES

Ackoff, Russel L. "Computer Obstructed Education," *Society for Management Information Systems,* 1973, pp. 108–114.

Ackoff, Russell L. "Towards a System of System Concepts," *Management Science,* July 1971, pp. 661–671.

Boulding, Kenneth. "General Systems Theory—The Skeleton of Science," *Management Science,* Apr 1956, pp. 197–208.

Churchman, C. W. *The Systems Approach.* New York: Dell Publishing Co., 1968.

Davis, Gordon. *Management Information Systems: Conceptual Foundations, Structure, and Development.* New York, McGraw-Hill, 1974.

Emery, F. E. (ed.). *Systems Thinking.* Baltimore: Penguin Books, 1969.

Johnson, Richard A., Fremont E. Kast, and James E. Rosenzweig. *The Theory and Management of Systems.* New York: McGraw-Hill, 1973.

Kast, Fremont E., and James E. Rosenzweig. *Organization and Management: A Systems Approach.* 2d ed. New York: McGraw-Hill, 1974.

Katz, Daniel, and Robert L. Kahn. "Organizations and the System Concept," *The Social Psychology of Organizations.* New York: Wiley, 1966, pp. 14–29.

Luchsinger, Vincent P., and V. Thomas Dock. *The Systems Approach: A Primer.* Dubuque, Iowa: Kendall/Hunt Publishing Co., 1975.

Nugent, Christopher E., and Thomas E. Vollman. "A Framework for the Systems Design Process," *Decision Sciences,* 1972, pp. 84–109.

Sayles, Leonard R., and Margaret K. Chandler. *Managing Large Systems.* New York: Harper & Row, 1971.

Von Bertalanffy, Ludwig. *General System Theory: Foundations, Development, Applications.* New York: George Braziller, Inc., 1968.

Weldon, R. J. "The Concept of a System," *The Journal of System Engineering* Vol. 1, No. 1 (Jan 1969), pp. 130–149.

Wendler, C. C. "What Are the Earmarks of Effective Total Systems," *Systems and Procedures Journal* XCII (July–Aug 1966), p. 30.

3 Concepts of Information and Organizations

This chapter has three primary thrusts. The first two consist of providing conceptual foundations in the areas of information and organizations. In the third section, the concepts of information and organizations are integrated.

INFORMATION

Information is the result of the capture and organization of data. In an information systems orientation, data become information when they are the basis upon which (hopefully) efficient and effective decisions can be made. That is, information is used to increase the probability that the right decision is made. The transforming of data into information is the primary function of an information system.

Using the systems concepts discussed in Chapter 2, a simple model of an information system can be illustrated as shown in Figure 3-1.

Value of Information

Few decisions are made with perfect information. Virtually all decisions are made with either some insufficiency of information required and/or an overload of information not required.

Figure 3-1. *Simple Model of an Information System*

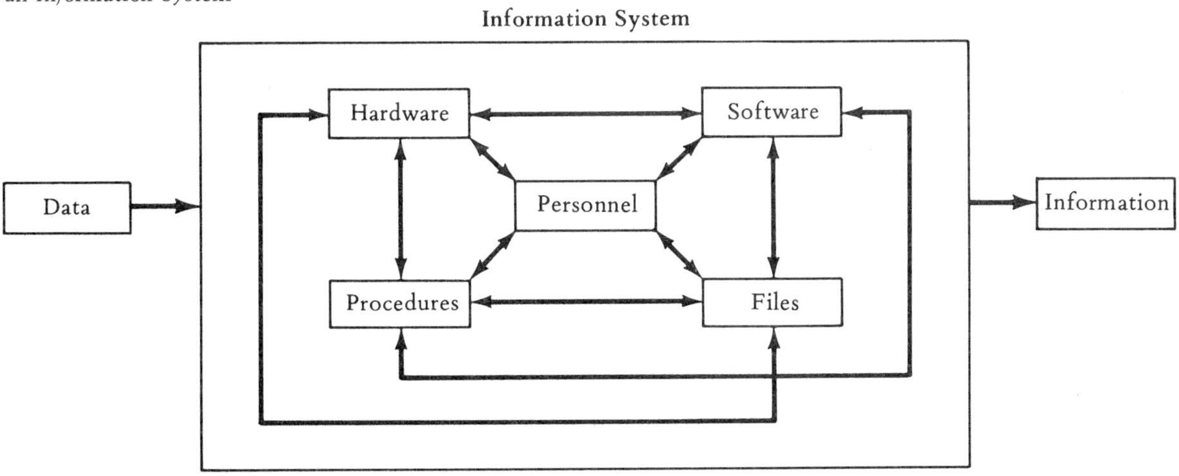

The value of information is a function of the effect it has on decision making. If additional information results in a better decision, the information has value. If additional information does not improve a decision, it has little or no value.

Consider a marketing manager who must determine whether or not to market a new product. The manager has tentatively decided against introducing the product but has asked the marketing research department to survey a sample of customers to determine interest in the product. The information resulting from this research indicates an 80 percent probability that the new product will be successful. Based on this information, the manager decides that perhaps introducing the product is a good idea. The additional information is, therefore, valuable; it results in the manager making what appears to be a better decision.

Suppose the manager wants to be more certain as to the success of the new product. He or she requests the marketing research department to conduct additional, more comprehensive research. The results of the second study indicate a 90 percent probability that the new product will be successful. This information confirms the manager's decision to introduce the new product. However, since this course of action had been determined earlier from previous information, the additional information has no particular value, other than that of increasing the confidence in a decision made previously.

Information can only improve a decision if it leads to a change in the decision. Accordingly, obtaining additional information is of value until it no longer changes a decision. In other words, information must have a "surprise" effect to be of value. It must tell the manager something important that he or she did not already know.

Note that, in the case of the marketing manager, even perfect information (i.e., such that the manager is 100 percent certain of success) would not have changed or improved the decision. The manager has already determined the correct course of action. In many situations, the optimum, or at least a reasonable, decision can be made without perfect information. This is fortunate, because obtaining perfect information for most decision making is impractical or impossible due to the factors listed in Table 3-1.

The problem with less than perfect information is that management can never be positive that additional information will have surprise value and result in a better decision. Accordingly, to the extent it is practical and possible, there is potential value in obtaining additional information. However, the question that should be asked is whether the cost of obtaining the additional information is justified.

Table 3-1. *Imperfections of Information for Decision Making*

Information Imperfection	*Illustrations*
Inability to accurately predict or control the future.	• Estimate customer demand for a new product. • Know the future availability of natural resources.
May be impractical or too costly to obtain information.	• Develop a profile of all airline passengers. • Maintain summary information on annual purchases made by customers of a supermarket chain.
Information is not available.	• Historical data on average customer balance in checking accounts. • Current employment of previous employees.
Unaware information exists.	• A company is unaware of a new government report on the labor market. • A company is unaware of a research project describing customer attitudes toward public transportation.
Information is in the wrong form.	• Sales data are summarized on an annual basis rather than monthly. • Inventory turnover statistics are available by item but not by customer demand.

Cost Justification of Information

There are two basic categories of value justification of information: those cases where the benefits can be calculated and those cases where they cannot. The benefits can be calculated when the effect on profitability of altering a decision can be determined. In this case, if the cost of additional information is less than the increase in profitability, obtaining the information is cost justified. For example, consider a department store chain that has a problem with lost or stolen credit cards being used to run up large bills at several stores before a stop-credit can be enforced. If the annual losses incurred by the company amount to $250,000 per year, an information system that can resolve the problem for a cost of $75,000 is quite justifiable.

In most situations, the benefits of additional information are not measurable. This makes it difficult to justify the cost of obtaining the information. An example of this situation is the current move of financial institutions to implement electronic funds transfer systems. Such systems minimize the need for checks, greatly expedite the flow of money between businesses, and improve the timeliness of financial information. Though these systems are very costly, little attempt has been made to cost justify them. There seems to be a general consensus that "faster is better," "new technology is good," and everyone (i.e., banks, stores, and customers) benefit.

The difficulty in assessing the benefits of additional information is a particularly perplexing dilemma since there is a tendency to seek or receive more information than is needed. Government agencies are often criticized for generating voluminous reports that have questionable value for decision making. Of course, since government agencies are nonprofit organizations, there is no thrust to cost justify information in terms of its contribution to profitability.

Even in business settings, cost justification of information becomes obscure. Suppose a loan officer at a bank requests information on the average checking account balance and the number of bad checks written for loan applicants. Certainly this information is meaningful when deciding whether or not to approve a loan. However, determining the number of bad loans avoided, and the consequent dollar savings, is difficult.

Information that cannot be explicitly cost justified must at least influence decisions if acquiring or processing it is to warrant consideration. Given that information requested does influence decisions, several factors that affect the cost of information require assessment. These factors are accuracy, timeliness, and reporting interval. They are discussed and illustrated in Table 3-2.

Framework for Information Systems

Information systems can be subdivided into four subcategories, or types. Each type of system is especially designed to perform a major function. The functions are referred to as transaction processing, information-providing, decision support, and programmed decision-making.

Table 3-2. *Factors That Affect the Cost of Information*

Factor	*Consideration*	*Example*
Accuracy	Increased accuracy generally means increased cost. Therefore, accuracy should only be increased to the point that it still influences decisions.	At the end of each business day, bank management must decide if the books are in balance. The bank can close a couple of hours earlier by allowing its balances to be off by a few dollars. A simple adjustment entry is much less costly than spending a few hundred dollars on salaries and computer time to locate a minor mistake.
Timeliness	In general, the more quickly information is provided after the occurrence of an event, the more costly is that information. More timely information can result in earlier problem identification and resolution.	Retailers like to identify slow-moving merchandise and put it on sale to move it before it becomes even less attractive to customers. A decrease in the demand for merchandise can be determined by keeping track of inventory turnover. However, such information can usually be a few days or weeks old without serious consequences.
Reporting Interval	In general, the more frequently information is updated and reported, the higher the cost. If information that affects decisions is changing quickly, the greater the need for shorter updating and reporting intervals.	An airline needs extremely short intervals for updating and reporting passenger reservations and changes in reservations to sell as many tickets as possible without overselling. However, the airline only needs to know the amount of passenger luggage to be carried once: just prior to take off.

Transaction-Processing Systems

Transaction-processing systems provide high-speed and accurate processing and record keeping of the basic clerical or operational processes of an organization. Examples of transaction-processing systems are the processing of sales transactions in a department store, checking accounts in a bank, or work orders in a manufacturing plant. Transaction-processing systems have traditionally been the most visible dimension of information systems since they often displace clerical personnel. The cost savings achieved by such displacement often provide a definitive basis for cost justification of a transaction-processing system.

Information-Providing Systems

Information-providing systems provide information necessary for decision making. This information is generally gleaned from transaction systems in the form of summary and exception reports. *Summary reports* are tabulations of detail by categories such as sales by store, by department, and by inventory item. *Exception reports* isolate detail or summary information that significantly deviates from normal. For example, an exception report may indicate large credit transactions, inventory items that are out of stock, or departmental payrolls that are exceeding budget.

Both summary and exception reporting are powerful dimensions of information systems. They highlight and isolate important detail information. This saves decision makers considerable time.

Decision-Support Systems

Decision-support systems go beyond the simple provision of information needed for decision making. They actually assist in the decision making. Decision-support systems include such things as statistical, mathematical, and simulation models that can be used by the decision maker to experiment with information in determining a course of action. For example, a simulation model can be used to forecast revenue for a new product.

Decision-support systems are relatively new, but they are gradually increasing in use. These systems are discussed in much greater detail later in this book.

Programmed Decision-Making Systems

Programmed decision making involves having a system make a decision rather than a person. Such decision making is preprogrammed according to very specific decision rules. For example, customer credit is refused if a customer has exceeded his or her credit. Before a system can be designed and implemented to program the computer for decision making, the decision-making process must be highly structured and defined.

Misconceptions about Information

Because information systems are able to process transactions, provide information for decision making, support decision making, and even make decisions, organizations have developed (often unrealistic) high expectations. Unfortunately, there is often a gap between what organizations expect of information systems and the performance of the systems.

This expectation gap results from six basic misconceptions about information systems and the role they play in organizations.[1] These misconceptions are identified and explained in Table 3-3.

ORGANIZATIONS

Organizations are human-designed and human-controlled systems. The entities of an organizational system are people, equipment, inventory, procedures, ideas, and so forth, arranged to interact to accomplish one or more objectives. For example, in a business organization, objectives typically include the production of certain goods or services, organization survival and growth, social responsibility, and, of course, profitability.

The most common way of illustrating an organization is to use an organization chart (see Figure 3-2). In this pyramidal structure, ultimate authority and responsibility reside at the top. Authority and responsibility flow down through the ever-widening succession of levels to the bottom of the organization.

An organization has both vertical and horizontal dimensions. The vertical dimensions are defined by the organization's departments or subunits. The horizontal dimensions are defined by the organizational layers or hierarchical structure.

Vertical Dimension

The vertical dimension of the organization performs the different functions required for the organization to operate (e.g., accounting, manufacturing, marketing, and research). The number of vertical

1. The misconceptions are based on the author's own work as well as the following articles: Russell L. Ackoff, "Management Misinformation Systems," *Management Science*, Dec 1967, pp. B147–156; and Chris Argyris, "Resistance to Rational Management Systems," *Innovation*, 1970, pp. 28–35.

Table 3-3. *Misconceptions about Information*

Misconception	*Explanation*
Computers can quickly and conveniently provide managers with information they desire.	Considerable organizational information is not stored on computer files. It must be collected before it can be processed. Once information is available, computer programs are required to access the information.
More information means better decisions.	Though most managers lack a good deal of the information they *should have,* many information systems provide them with an overabundance of irrelevant information they *don't need.* Many managers receive voluminous reports that simply sit unused on shelves because the reports are not relevant to decision-making processes.
Managers need the information they want.	For managers to know what information they need, they must be aware of each decision they should and do make. They also must define the process and variables required to make each decision. This situation is seldom the case. Consequently, managers tend to "play it safe" and ask for more information than they really need.

dimensions in an organization is a function of the differentiation (i.e., division of labor) and specialization required to perform the various tasks in the organization.

In a small one-owner proprietorship, the owner may perform all of the various tasks associated with sales, accounting, purchasing, inventory, and so forth. However, the volume and complexity of a large corporation require that the various tasks be "divided and conquered." Different departments are established to manage different functions, and people specialize in performing the tasks associated with particular functions. For example, an organization's accounting process becomes so voluminous and complex that data processing equipment is necessary. The accounting personnel do not have the technical skills needed to program and

Table 3-3 *(cont'd)*

Misconception	*Explanation*
Give managers the information they need and their decision making will improve.	Managers may not be aware of the best way to use information to make the best decision. For example, several managers given the same information make different decisions, and some of the decisions are better than others. Since all managers had the same information, the better decisions are a result of the decision makers performance, not of the information.
Managers do not have to understand how information systems work.	Managers may lack confidence in something they do not understand and consequently not use it. Also, managers are less apt to recognize logic problems in the information generated by an information system if they do not understand how the information is processed.
Managers always welcome and encourage imporved information for decision making.	Often, improved information reveals that an operation is poorly managed (e.g., incompetence, missed opportunities, coverups). Improved information about an operation can be threatening and, therefore, resisted by managers.

operate computer equipment; consequently, the data processing function is differentiated into another department having personnel who specialize in data processing skills. Even within the data processing department, further departmentalization and specialization are required (e.g., systems analysis, programming, and computer operations).

Horizontal Dimension

The horizontal dimension of an organization can be broadly categorized as top, middle, and operating management. In Figure 3-2, top management includes the board of directors, president, and vice presidents. Middle management includes the controller

Figure 3-2. *An Organization Chart*

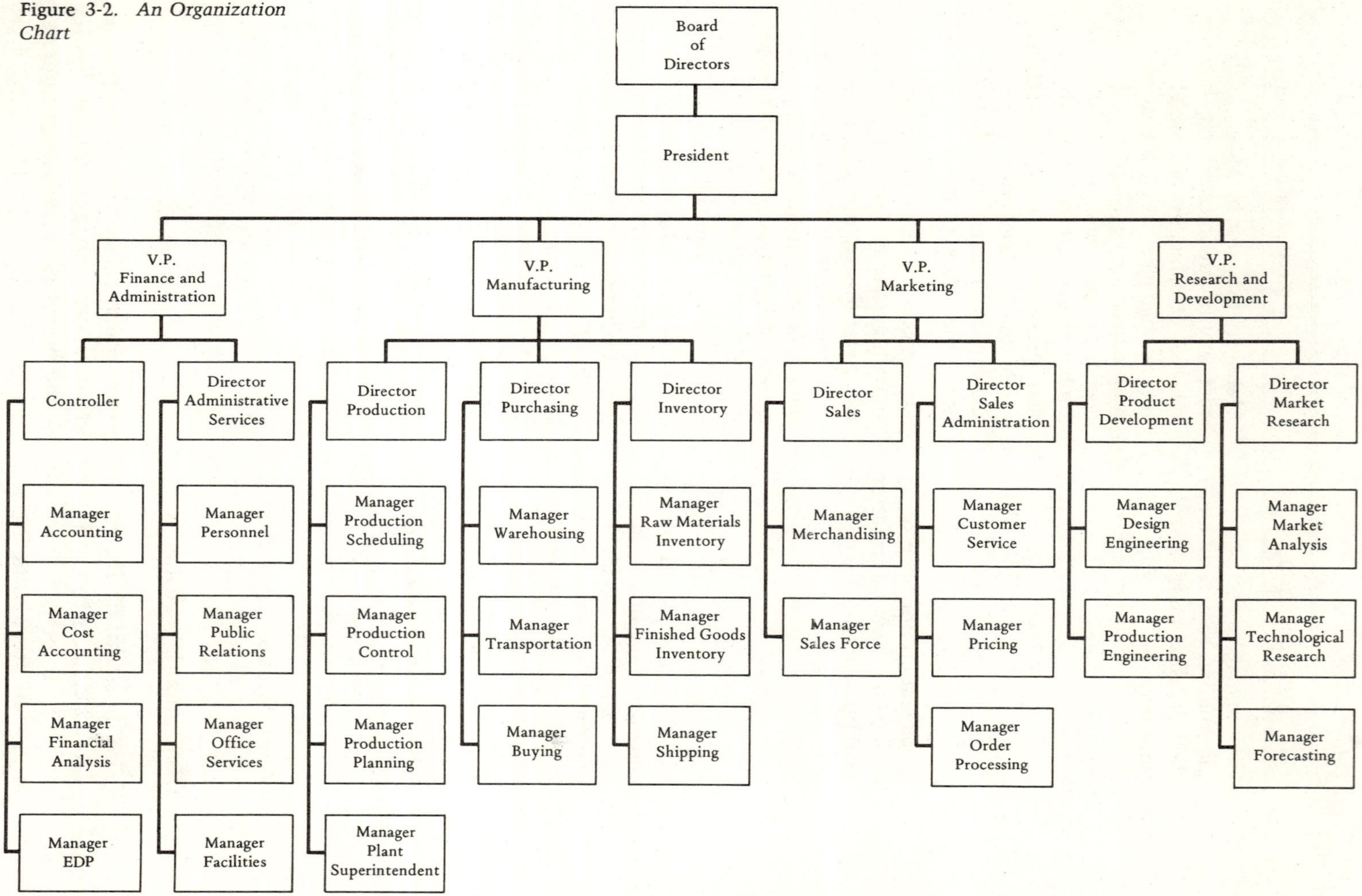

Table 3-4. *Horizontal Dimensions of an Organization*

Management Level	*Function*
Top Management	• Defines overall organizational objectives and formulates strategic plans and policies to achieve objectives. Allocates the required financial and human resources among the vertical dimensions of the organization.
Middle Management	• Develops, directs, and controls departmental objectives, plans, and policies consistent with overall organizational objectives and available resources. Integrates and coordinates the vertical dimensions of the organization.
Operating Management	• Supervises the daily production of goods and/or services consistent with performance criteria established by middle management.

and various directors. Operating management includes the remaining management positions. The major functions of these levels are given in Table 3-4.

Management of Organizations

The concept of organization implies rational, integrated, and coordinated interaction of the organizational subsystems and entities. However, inevitable conflicts and discord arise. These can be dysfunctional to the performance and even the survival of an organization. Management is required to maintain harmony of the (often diverse) dimensions of the organization to insure a viable and workable system.

Since an organization is a system, it depends on its environment to supply it with inputs (material, energy, and information) and to consume its outputs (goods and/or services). Since the demand for goods and services changes with time, an organization must be sensitive to market conditions to insure consumption of its output. An organization must also be concerned about the continued availability of necessary inputs (e.g., natural resources). Therefore, besides addressing issues internal to the organization, management must be concerned with co-aligning organizational processes with the inputs available from and the outputs desired by the organization's environment.

Approaches to Management

Over the years, various theories of management have been proposed. Many people have attempted to determine the "best way" to manage organizations. Broadly speaking, existing management theories form a continuum between two extreme positions with respect to management: the classical approach and the human relations approach. The primary management concepts within each approach are listed in Table 3-5.

The classical and human relations approaches to management are obviously opposing in perspective. Both approaches have been used with success in some organizations and have been unsuccessful in others. This has understandably created confusion and debate as to which is the better way to manage organizations.

Contingency Theory of Management

Actually, there is not a single best way to manage all organizations. Rather, the best management approach is *contingent* upon the characteristics of the organization being managed. This concept is referred to as the *contingency theory* of management. It attempts to recognize organizational differences and propose management approaches that "best fit" those organizational differences.

Table 3-5. *Contrasting Approaches to Management*

Classical Approach	*Human Relations Approach*
Formal	Informal
Highly Structured	Flexibly Structured
Centralized Decision Making	Decentralized Decision Making
Dictatorial Management	Participative Management
Programmed Procedures	Dynamic Procedures
Control Orientation	Autonomy Orientation
Process Orientation	Personnel Orientation

For example, the organizational characteristics of an assembly line in a manufacturing plant are different than the organizational characteristics of a group of research scientists. An assembly line can be characterized as routine and repetitive whereas research is creative and diverse. To be effective, each group should be managed differently. The management challenge becomes one of systematically identifying the relevant differentiating characteristics and then applying the appropriate management techniques to each group.

Technology and Environment

The contingency theory of management has established that the technology and/or the environment of an organization are the key variables that characterize the organization. *Technology* is defined as the human and mechanical process by which an organization produces its goods and/or services. *Environment* is defined as all entities or factors external to the organization. An organization's environment includes such things as customers, competitors, and government regulations.

The classification of systems defined in Chapter 2 can be used to characterize an organization. That is, the technology and environment of an organization establish the organization's position on a continuum ranging from closed/stable/mechanistic to open/adaptive/organic.

A closed/stable/mechanistic organization has a predominantly routine and predictable situation. Productivity is a major objective. Technology and environmental forces are relatively stable and certain. Consequently, decision making tends to be straightforward and programmable. An example of a closed/stable/mechanistic organization is the can manufacturing industry. It has used a relatively stable technology and encountered a relatively stable environment for over twenty years.

Alternatively, an organization that is open/adaptive/organic is involved with other than routine activities. Creativity and innovation are required. Technology and environmental forces tend to be volatile, presenting considerable uncertainty. Problems encountered are complex, presenting conflict and unclear courses of action. Decision-making processes are characterized as risky, and few optimum decisions are possible. An example of an open/adaptive/organic organization is the electronic calculator industry. It has had an extremely dynamic technology and environment.

Few organizations or organizational subsystems are polarized as either closed/stable/mechanistic or open/adaptive/ organic. Rather, organizations generally tend, in varying degrees, to be one way or the other. Also, an organization's position on the continuum will occasionally vary, perhaps extremely, with time. This occurs when an organization encounters changing technologies and/or environments. For example, the energy crisis considerably impacted the automotive marketplace. Consequently, automobile manufacturers have had to cope with a less predictable environment as they contemplate future product offerings (e.g., luxury cars vs fuel-efficient cars).

The realization that organizations vary in form from closed/stable/mechanistic to open/adaptive/organic to accommodate their technology and environment provides the basic framework for contingently applying managerial approaches. A management style aligned with the classical approach tends to be effective in relatively closed/stable/mechanistic organizations. Management can operate in a structured, centralized, and process-oriented mode. Conversely, the human relations management approach tends to be effective in open/adaptive/organic situations where employees are coping with greater uncertainty and complexity. Management must provide a more autonomous and participative setting for what is generally a more sophisticated level of staff. In fact, in most open/adaptive/organic settings, staff members generally have more expertise in special areas than their managers. Figure 3-3 illustrates the relationship between organiza-

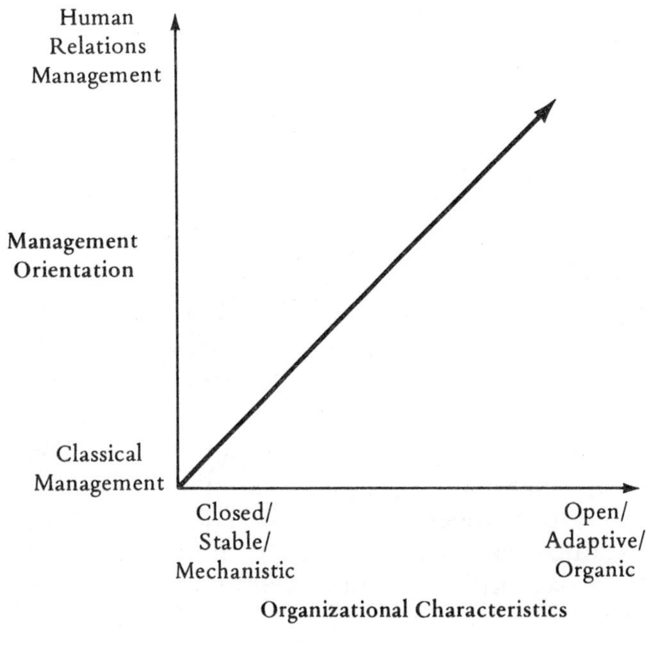

Figure 3-3. *Relationship between Organizational Characteristics and Management Orientation*

tions or organizational subsystems and the appropriate management orientation.

The contingency view of management and organizations can also be applied to both the horizontal and vertical dimensions within an organization. From a horizontal perspective, the lower levels of the organization are generally closed/stable/mechanistic in orientation. The orientation of the organization becomes increasingly open/adaptive/organic as one moves from the operating level to top management.

From a vertical perspective, the accounting department provides an excellent example of a process that is inherently structured and, therefore, leans toward a closed/stable/mechanistic orientation. Conversely, an advertising department, where creativity and spontaneity are essential, represents the opposite end of the spectrum. Figure 3-4 offers a contingency view of the relationship between the horizontal and vertical dimensions of a hypothetical organization.

Organizational Considerations for Information Systems

Even limited experience reveals that computer technology is highly structured. Computer programming requires great precision and attention to details. Accordingly, computer-based information

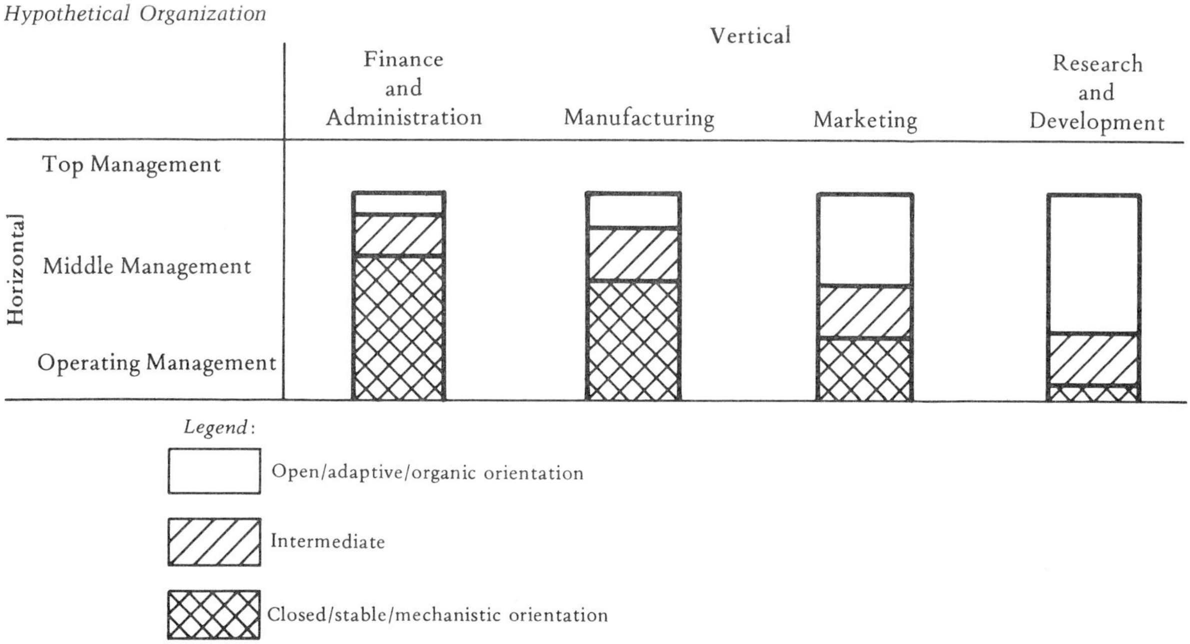

Figure 3-4. *Contingency View of the Vertical and Horizontal Dimensions of a Hypothetical Organization*

processing is most commonly applied in the closed/stable/mechanistic dimensions of organizations. Figure 3-4 indicates that there are both horizontal and vertical organizational dimensions that readily lend themselves to computer-based information systems.

Horizontal Dimension

In the horizontal dimension, the information needs of operating management are more easily addressed than the needs of middle and top management. Ostensibly, at progressively higher organizational levels, it becomes increasingly difficult to accommodate the decision-making processes with the desired information.

Due to the greater structure and predictability at lower organizational levels, decision making tends to be more straightforward and computational. Most of the necessary information is internal to the organization. Usually, the information is complete. Most of the lower-level information requirements are accommodated by information generated from the transaction-processing systems.

Higher-level management is working with less definition, structure, and routineness in decision making. In fact, much of decision making is involved with issues external to the organization. Accordingly, information external to the organization plays a key role in upper-management decision making. Unfortunately, external information tends to be voluminous, unstructured, volatile, and inaccessible. Consequently, it is impossible to provide to this level of management all the information necessary for decision making. The lack of definition, structure, and routineness in upper-management decision making, combined with the external and incomplete qualities of the necessary information, tends to make upper-management decisions highly judgmental.

A profile of the basic differences in both information and decision making at the various management levels is given in Table 3-6.

Vertical Dimension

The closed/stable/mechanistic and open/adaptive/organic tendencies of the vertical, or functional, dimension vary (review Figure 3-4). The finance and administrative functions, which are common to all organizations, are usually the most closed/stable/mechanistic. Consequently, accounting and administrative functions are generally the first to be included in computer-based information systems developed and/or used by an organization.

Table 3-6. *Characteristics of Decision Making and Information at Different Management Levels*

	Management Level		
	Operating	*Middle*	*Top*
Characteristics of Decision Making	Computational/Objective →		Judgmental/Subjective
Examples	Inventory Reordering Production Scheduling Credit Approval	Short-Term Forecasting Budget Preparation Capacity Planning	New Product Planning Location of New Factory Mergers and Acquisitions
Characteristics of Information	Internal/Complete →		External/Incomplete
Examples	Sales Orders Production Requirements Customer Credit Status	Sales Analysis Budget Analysis Production Summaries	Market Conditions Industry Forecasts Government Regulations

The remaining vertical dimensions vary with organizations. Their characters are determined by the goods or services an organization produces, and the technologies associated with producing those goods or services. For example, hospitals, banks, retailers, manufacturers, and universities differ in the goods and services they deliver and consequently in their vertical structures. They all have finance and administrative functions, but their "mainstream" ("core") activities vary. However, the mainstream activities do share certain commonalities. Basically, all mainstream activities involve some form of the following activities:

- Forecasting demand
- Scheduling or allocating resources
- Inventorial activities
- Control

Illustrations of these four activities in the contexts of different organizations are provided in Table 3-7.

The mainstream activities of some organizations are more closed/stable/mechanistic than those of others. Consequently, certain industries have experienced earlier and more far-reaching success in developing computer-based information systems. Banking and manufacturing are examples of industries that are considered ideal for computer-based information systems. Nearly half of the computer equipment installed today is utilized by the banking and manufacturing industries. Such success has materialized more slowly in industries such as medical care and retailing, which

Table 3-7. *Activities Common to Mainstream Processes of Different Organizations*

Organization	*Mainstream Process*	*Forecasting*	*Scheduling*	*Inventory*	*Control*
Manufacturer	Manufacturing	Demand for existing and new products Future availability of raw materials and labor	Work orders Equipment Labor	Finished goods Raw materials Labor Equipment	Performance of product Market acceptance of product Productivity assessment
University	Instruction Research Community service	Future demand for higher education Future availability of faculty	Students Classrooms Equipment Faculty	Students Classrooms Equipment Faculty	Placement of students Students' performance on exams Quality of teaching Research activity Resource utilization
Airline	Transportation of passengers and freight	Future demand for air travel Future supply of aircraft	Passengers Freight Flights Aircraft Personnel	Repair parts Aircraft Personnel	Safety record Customer satisfaction Flight occupancy
Bank	Checking and savings accounts Installment and mortgage loans	Future supply of money in checking and savings accounts Future demand for installment and mortgage loans	Operational personnel Cash flow Investments	Dollars Personnel	Market share Loan defaults Liquidity
Hospital	Health care	Future need for health care Future availability of hospital facilities and physicians	Patients Rooms Equipment Physicians	Medical supplies Patients Rooms Equipment Physicians	Patient care Facility utilization

combined, utilize less than 5% of currently installed computer equipment.[2] However, generalizations should not be made about industries as to their closed/stable/mechanistic orientation. For example, in manufacturing, large batch and mass production is more closed/stable/mechanistic than production in unit and small batch. The latter is involved with tailoring production to customer requirements, production of prototypes, and fabrication, which are more open/adaptive/organic processes. Therefore, in designing an

2. Jerome Kanter, *Management-Oriented Management Information Systems,* 1977, p. 40.

information system, the particular organization and organizational processes within the organization must be considered on an individual basis. The more closed/stable/mechanistic a particular process, the more the information system can be oriented toward routine transaction processing and programmed decision making. The more open/adaptive/organic a process, the more the information system should be oriented toward providing, to the extent possible, information for decision making and decision-support capabilities.

MANAGEMENT INFORMATION SYSTEMS

From the discussions of vertical and horizontal information systems, it can be concluded that an organization has several information systems. For example, there are financial information systems that have subsystems consisting of general ledger, accounts receivable, accounts payable, cost accounting, and payroll/personnel. There are manufacturing information systems that have subsystems consisting of production scheduling, inventory, and material requirements. The various information systems address, to varying degrees, the information and decision-making needs of operating, middle, and top management.

The vertical and horizontal information systems can be envisioned as a federation of information subsystems that constitutes an overall organizational information system called a *management information system* (MIS). MIS is a somewhat overwhelming concept. Considerable organizational and information systems knowledge and experience are required to totally appreciate it. When attempting to build an effective and efficient MIS, all of the various information subsystems are designed and developed in the context of the total MIS. Where appropriate, the various subsystems are interfaced to supply information to one another. For example, the payroll/personnel subsystem requires information from, and supplies information to, every other vertical and horizontal dimension of the organization. Sales information is used to determine production schedules; production schedules are used to determine raw materials requirements; and so forth. An overall conceptualization of an organizational MIS is given in Figure 3-5.

A more comprehensive treatment of MIS is presented in the final section of this text, "Advanced Systems Analysis." To prepare the student for that material, it is critical to first develop

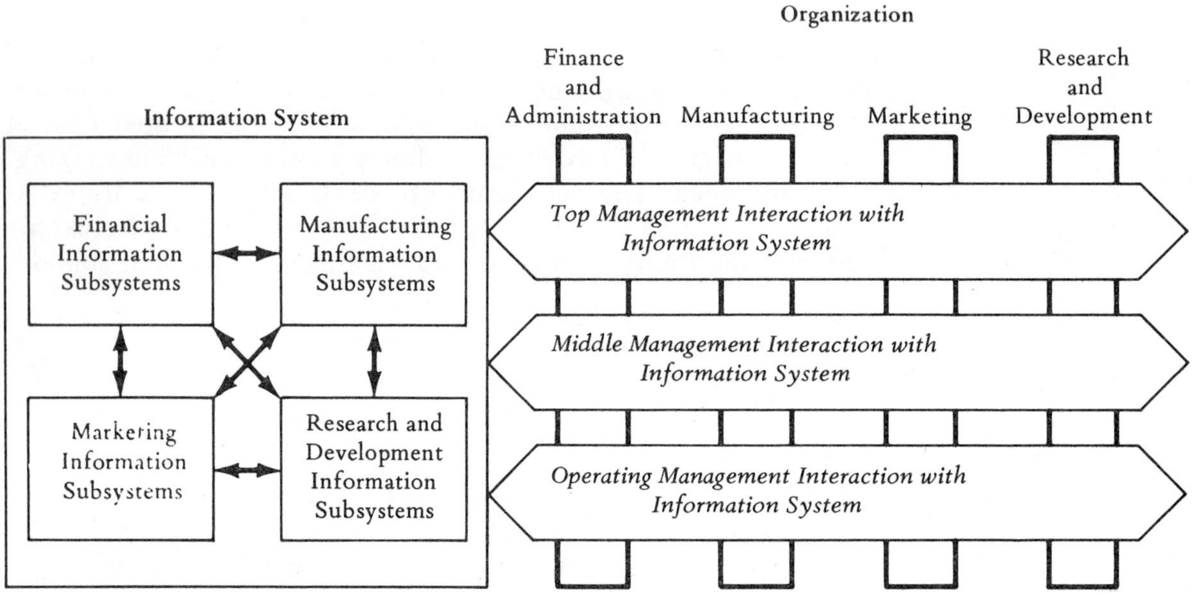

Figure 3-5. *Conceptualization of a Management Information System*

systems analysis skills in working with smaller, more manageable, information subsystems. Accordingly, prior to the final section, the text concentrates on systems analysis at the subsystem level. However, the student is encouraged to keep in mind that some integration of these subsystems is necessary to achieve an overall organizational MIS.

SUMMARY

Information is the primary basis for decision making. The value of information is a function of the effect it has on increasing the probability that the right decision is made. Virtually all decisions are made on the basis of imperfect information. However, additional information should be sought, only to the point that it is likely to influence a decision.

The four basic categories of information systems are: transaction processing, information-providing, decision-support, and programmed decision making.

There are several misconceptions about what information can do for organizations. These misconceptions are based on erroneous assumptions about the relationship between management decision making and information.

Organizations are human-designed and human-controlled systems. They have both vertical and horizontal dimensions, which vary between closed/stable/mechanistic and open/adaptive/organic. The more closed/stable/mechanistic a process, the more readily it can be supported by a computer-based information system. However, both improvements in technology and more advanced analytical skills are continually facilitating the development of new computer-based information systems applications in open/adaptive/organic environments.

An organization has a federation of information subsystems that constitutes a management information system (MIS). Proper interfacing of these subsystems is critical to proper integration of the collection, processing, and providing of information to the organization.

EXERCISES

1. Why is it often difficult to cost justify information? If information cannot be cost justified, what should be the primary criteria in assessing the utility of providing it?

2. Identify and discuss the major factors that affect the cost of information.

3. Categorize the following functions as pertaining to transaction processing, information providing, programmed decision making, or decision support.
   a. Credit is automatically rejected because a customer has a balance past due.
   b. Management is provided with a sales analysis by department.
   c. An online financial terminal in a savings and loan captures data on deposits and withdrawals.
   d. A simulation model is used to forecast profitability for a company, based on projected sales.

4. Discuss the major misconceptions about information systems and the role they play in organizations.

5. Discuss the differences between the vertical and horizontal dimensions of an organization.

6. Discuss the effect of an organization's technology and environment as it pertains to the organization being closed/stable/mechanistic or open/adaptive/organic.

7. Identify the following organizational processes as being relatively closed/stable/mechanistic or relatively open/adaptive/organic.
   a. Processing checking accounts in a bank
   b. Payroll processing
   c. Advertising
   d. Credit management
   e. Diagnosing medical treatment for a heart patient
   f. Determining whether to merge with another corporation
   g. Strategic planning
   h. Reordering inventory
   i. Designing an information system
8. Discuss the difference between a single information system and an MIS.

SELECTED REFERENCES

Aron, Joel D. "Information Systems in Perspective," *Computing Surveys,* Dec 1969, pp. 213–236.

Ackoff, Russell L. "Management Misinformation Systems," *Management Science,* Dec 1967, pp. B147–156.

Andrus, Roman R. "Approaches to Information Evaluation," *MSU Business Topics,* Summer 1971, pp. 40–46.

Argyris, Chris. "Resistance to Rational Management Systems," *Innovation,* No. 10 (1970), pp. 28–35.

Bowers, David G., and Stanley E. Seashore. "Predicting Organizational Effectiveness with a Four Factor Theory of Leadership," *Administrative Science Quarterly* Vol. II, No. 2 (1966), pp. 238–263.

Chervany, Norman L., and Gary W. Dickson. "An Experimental Evaluation of Information Overload in a Production Environment," *Management Sciences,* June 1974, pp. 1335–1344.

Davis, Gordon B. *Management Information Systems: Conceptual Foundations, Structure, and Development.* New York: McGraw-Hill, 1974.

Dickson, Gary W., and John K. Simmons. "The Behavioral Side of MIS," *Business Horizons,* Aug 1970, pp. 59–71.

Etz, D. V. "The Marginal Utility of Information," *Datamation,* Aug 1965, pp. 41–44.

Feltham, Gerald A. "The Value of Information," *The Accounting Review,* Oct 1968, pp. 684–694.

Fiedler, Fred E. "Validation and Extension of the Contingency Model of Leadership: A Review of Empirical Findings," *Psychology Bulletin* Vol. 76, No. 2 (1971), pp. 128–148.

Galbraith, Jay R. "Organization Design: An Information Processing View," *TIMS Interfaces,* May 1974, pp. 28–36.

Gregory, Robert H., and Thomas V. V. Atwater, Jr. "Cost and Value of Management Information as Functions of Age," *Accounting Research* Vol. 8, No. 1 (Jan 1957), pp. 42–70.

Hirsch, Rudolph E. "The Value of Information," *The Journal of Accountancy,* June 1968, pp. 41–45.

Hunt, Raymond G. "Technology and Organization," *Academy of Management,* Sept 1970, pp. 235–252.

Johnson, Richard A.; Fremont E. Kast; and James E. Rosenzweig. *The Theory and Management of Systems.* New York: McGraw-Hill, 1973.

Kanter, Jerome. *Management-Oriented Management Information Systems.* 2d ed. Englewood Cliffs, N.J.: Prentice-Hall, 1977.

Kast, Fremont E., and James E. Rosenzweig. *Contingency Views of Organization and Management.* Chicago: Science Research Associates, 1973.

Kast, Fremont E., and James E. Rosenzweig. *Organization and Management: A Systems Approach.* 2d ed. New York: McGraw-Hill, 1974, pp. 510–511.

Katz, Daniel, and Robert L. Kahn. "Organizations and the System Concept," in *The Social Psychology of Organizations.* New York: Wiley, 1966, pp. 14–29.

Lawrence, Paul R., and Jay W. Lorsch. *Organization and Environment.* Homewood, Ill.: Irwin, 1967.

Lawrence, Paul R., and Jay W. Lorsch. *Organizational Structure and Design.* Homewood, Ill.: Irwin, 1970.

Perrow, Charles. "A Framework for the Comparative Analysis of Organizations," *American Sociological Review,* Apr 1967, pp. 194–208.

Pugh, D. S.; D. J. Hickson; and C. R. Hinings. *Writers on Organization.* 2d ed. Harmondsworth, England: Penguin Books, 1971.

Terreberry, Shirley. "The Evolution of Organizational Environments," *Administrative Science Quarterly,* Mar 1968, pp. 490–613.

Thompson, James D. *Organizations in Action.* New York: McGraw-Hill, 1967.

Wetherbe, James C., and Carlton J. Whitehead. "A Contingency View of Managing the Data Processing Organization," *MIS Quarterly,* Mar 1977, pp. 19–25.

Woodward, Joan. *Management and Technology.* London: Oxford University Press, 1965.

4 Concepts of Data Processing

Hardware

Input/Output Devices
Secondary Storage Devices
Central Processing Unit (CPU)
Data Entry and Reporting

Software

Systems Programs
Application Programs

Files

Accessing Files
Database Management Systems

Procedures

Systems Development Procedures
Operating Procedures

Personnel

Position Descriptions
Organization and Management of Data Processing Personnel

Summary

Exercises

Selected References

This chapter presents basic concepts of data processing as they pertain to computer-based information systems. The orientation of this material is fundamental and well within the grasp of students with limited data processing exposure. For students with more extensive data processing backgrounds, the material is both a good review and a view of data processing within a systems framework.

In Chapter 3 a model of an information system was illustrated as shown in Figure 4-1. In this chapter the five subsystems—hardware, software, files, procedures, and personnel—are defined and discussed using the systems concepts developed in Chapter 2.

HARDWARE

The core of data processing hardware consists of the computer system itself. The major entities of a computer system are:

- Input devices
- Output devices
- Secondary storage devices
- Central processing unit (CPU)

Figure 4-1. *An Information System*

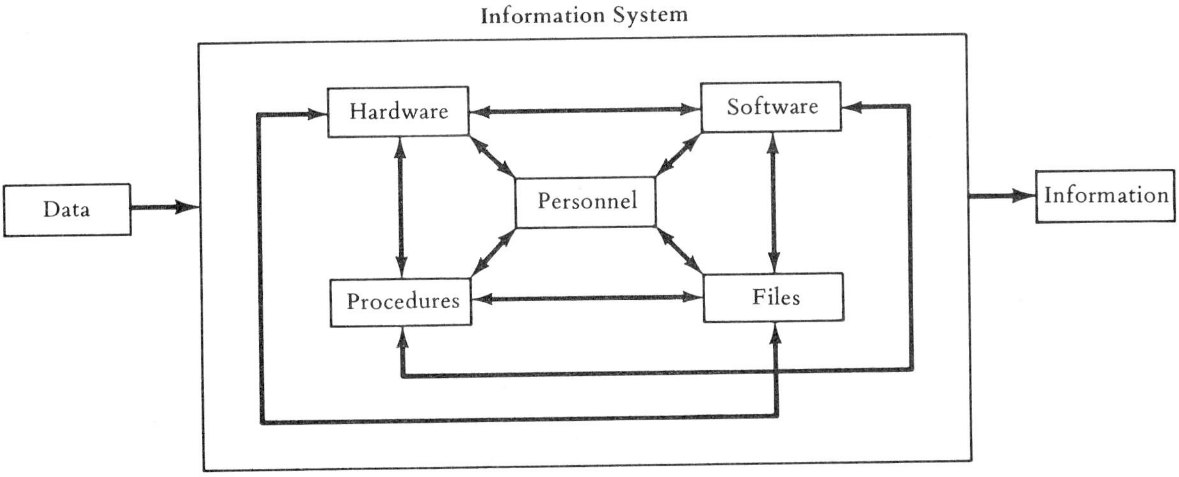

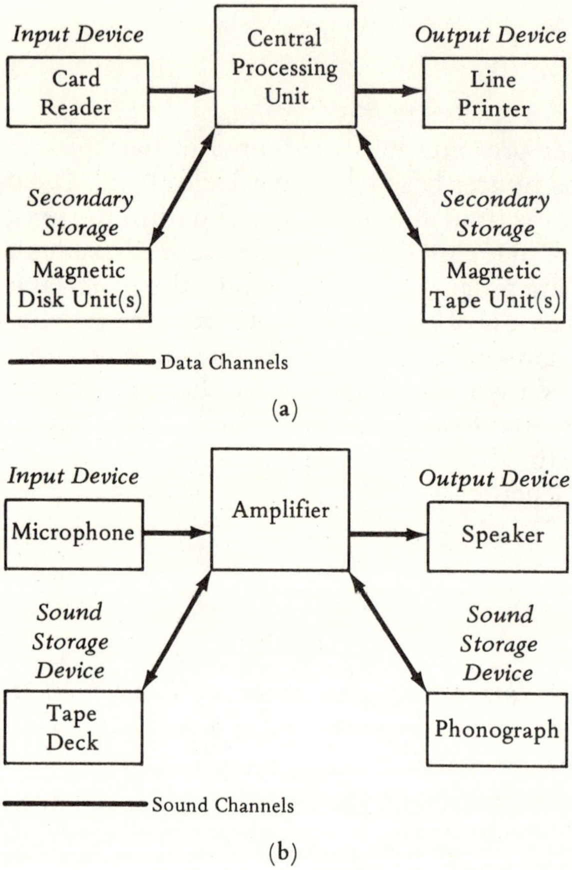

Figure 4-2. *(a) Model of a Computer System. (b) Model of a Stereo System*

The traditional input and output devices of computer systems are card readers and line printers, respectively. The most common secondary storage devices are magnetic disk unit(s) and magnetic tape unit(s). Data channels are used to connect these devices to the CPU. A model of a computer system is shown in Figure 4-2*a*.

For the reader unfamiliar with computer hardware, there is a remarkable resemblance between the components of a computer system and the components of a stereo system (Figure 4-2*b*). Since most if not all readers are familiar with stereo equipment, it is helpful to draw analogies on a component basis between stereo systems and computer systems. Such analogies are used in the following discussion of computer system components.

Input/Output Devices

A wide variety of input and output devices may be used on a computer system. Commonly used devices are mentioned below.

Input Devices

The card reader is used to enter data into a computer system. A card reader is therefore analogous to the microphone on a stereo system. Each device facilitates inputs to its respective system. Card readers generally operate at speeds ranging from 100 to 2000 cards per minute. Other commonly used input devices include computer terminals (e.g., hardcopy, typewriterlike and cathode ray tube terminals), key-to-disk and key-to-tape data entry systems, optical scanners, and magnetic ink character readers (MICRs).

Output Devices

The output of computer processing is information. The most common form of output is a report generated on a printer. The output from a stereo system is sounds generated from speakers. Therefore, the printer is to the computer system what the speakers are to the stereo system.

Printers generally operate at speeds ranging from 300 to 2000 lines per minute. Other output devices include computer terminals, graphic plotters, and computer output microfilm (COM) units.

Interactive Devices

Most computer terminals are both input and output devices. They are generally referred to as "interactive" devices, since they can both transmit and receive data in a "conversational" mode. The interactive capability of such terminals has made them extremely popular. Most computer systems include interactive terminals.

Secondary Storage Devices

The most common data storage media on computer systems are magnetic tapes and magnetic disks. The roles and operation of magnetic tapes and disks on a computer system are analogous to the roles and operation of tape recordings and phonograph records on a stereo system. A major difference is that on the computer system data is stored, but on a stereo system sounds are stored.

Magnetic Tape

A magnetic tape unit is referred to as a *sequential access device.* Data are sequentially stored on and retrieved from a magnetic tape in much the same manner as recordings are sequentially recorded on or played back from a tape recording. On a stereo tape, the preceding nine recordings have to be physically bypassed to get to the tenth recording. Similarly, on a magnetic tape, the preceding 1999 payroll records have to be physically bypassed to access payroll record 2000.

A full 2400-foot reel of magnetic tape can generally store over 30 million characters (i.e., numbers, letters, or special characters). A magnetic tape unit can read or write a full reel of magnetic tape in five to ten minutes.

Magnetic Disk

A magnetic disk unit is referred to as a *direct access device.* It has read/write heads that can both read and write data on the surface areas of platter-shaped magnetic disks (see Figure 4-3). As depicted in Figure 4-3, access arms are used to shift the read/write heads back and forth over the surface areas of the rotating disks.

If it is temporarily assumed that a record player can record sounds on a phonograph record as well as play them back, a record player and a disk unit become quite similar. The ability to shift the read/write heads on the disk unit directly over a particular track is analogous to the capability of positioning the needle of a record

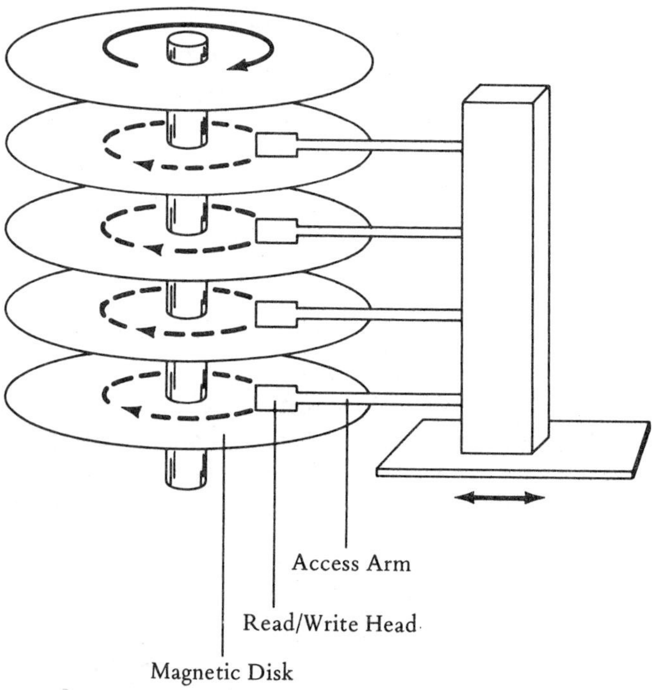

Figure 4-3. *A Magnetic Disk Drive*

player directly over a particular recording. It is also possible to sequentially progress to a particular location on either device.

The time associated with the shifting of a read/write head over a track of data is referred to as *seek time.* Once the read/write head is positioned over the desired track, it must wait for the desired data to rotate (or spin) under the read/write head. This is referred to as *rotational delay.* Therefore, the time required to access data on a disk is the sum of the seek time and the rotational delay.

Magnetic disk technology is superior in both storage capacity and access speed to magnetic tape technology. However, magnetic tape is more economical. A reel of magnetic tape costs approximately $15; a disk pack costs from $400 to $2000. Magnetic tape units are also less costly than magentic disk units. However, disk processing costs have dropped dramatically during recent years. Consequently, disk processing now dominates the secondary storage market.

Central Processing Unit (CPU)

The CPU of a computer system is the "heart" of the computer. It is similar to the amplifier of a stereo system in that both have consoles and both are connected to all systems devices. However, they are quite different in function and operation.

A CPU is comprised of a control unit, an arithmetic/logic unit, and a primary storage unit. These components and their interaction are illustrated in Figure 4-4. The control unit monitors and regulates the overall operation of the system and the execution of computer programs. The actual arithmetic and logical processing occur in the arithmetic/logic unit. The primary storage unit, or computer memory, is used to store data and instructions currently

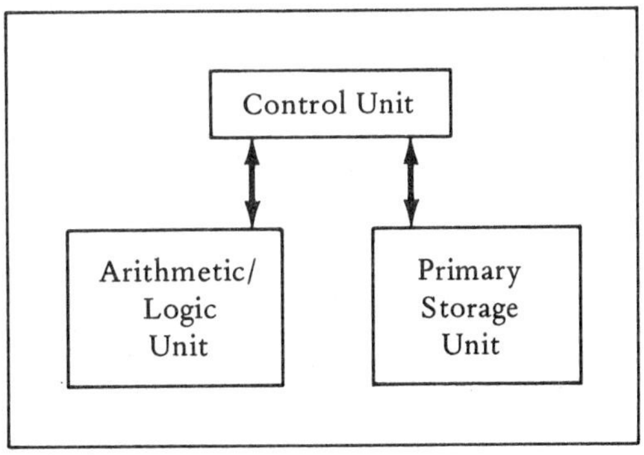

Figure 4-4. *Central Processing Unit (CPU)*

in use by the CPU. The storage capacity of a primary storage unit generally ranges from 8000 characters on a small system to several million characters on a large system.

The speed at which CPUs operate can generally be measured in terms of *nanoseconds,* or thousand millionths of a second. Viewed as a relative unit of measure in time perspective, there are as many nonoseconds in a second as there are seconds in thirty years. The CPU of a computer system is therefore operating at significantly faster speeds than any other component of the system (e.g., even disk drives are several thousand times slower than a CPU). Consequently, the average CPU can concurrently run several computer programs accessing various input, output, and secondary storage devices and still have unused (idle) capacity.

Data Entry and Reporting

Computers are quite efficient at processing data. However, the logistics associated with data entry and reporting represent two of the major difficulties and expenses of data processing. The two basic modes of data entry and reporting are batch processing and online processing.

Batch Processing

Batch processing is a computer processing technique whereby a number of similar input items are grouped together for processing during the same computer run. It involves converting transactions into machine-readable form prior to inputting them to the computer system. The punched card is the most commonly used medium for batch processing, but it is being replaced by key-to-disk units in many systems.

Keypunching of customer orders and daily entry of them into the computer for processing is an example of batch processing. A batch update process using punched cards is shown in Figure 4-5*a*. Other devices commonly used for data entry in batch processing made include key-to-tape, optical scanning, and MICR devices.

Reporting in a batch-processing environment generally consists of printing and disseminating reports on a periodic basis. For example, weekly printouts are prepared to report outstanding customer orders and then distributed to the various offices that need the information.

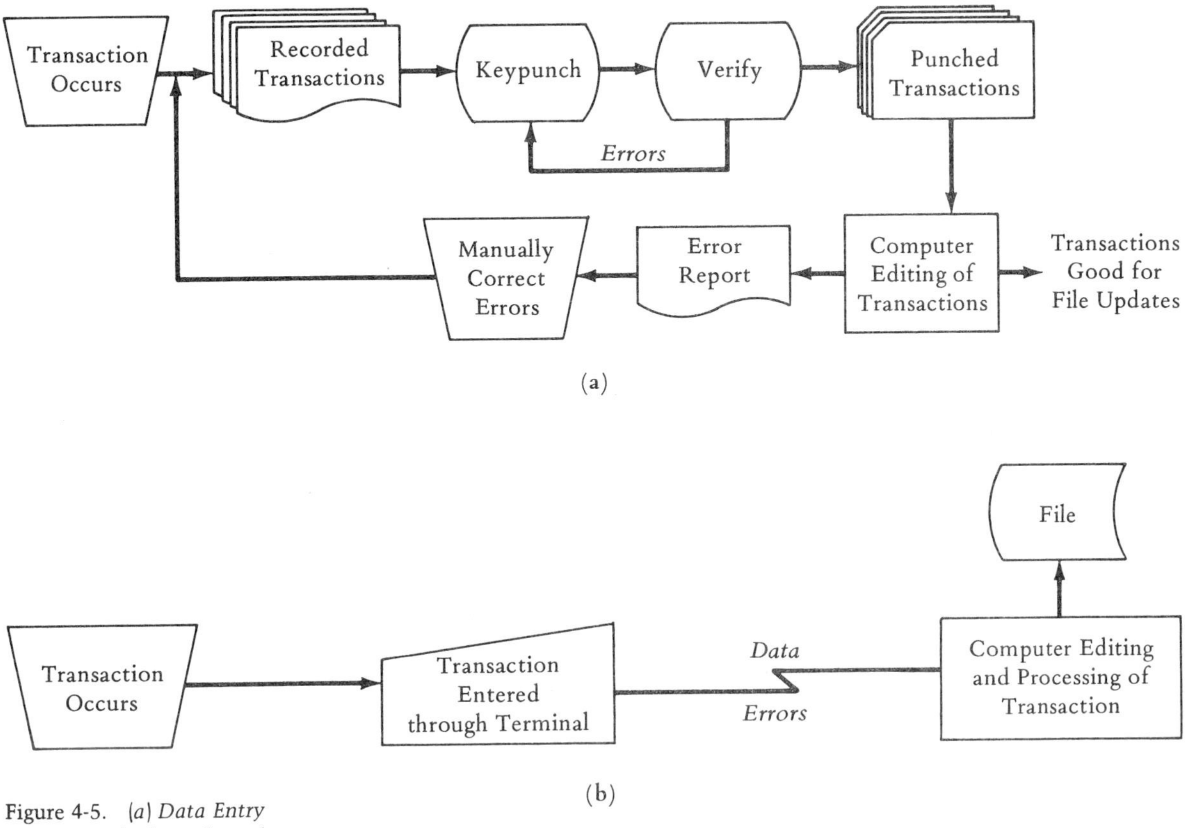

Figure 4-5. *(a) Data Entry Using Punched Cards and Batch Processing. (b) Data Entry Using Online Technology*

Online Processing

Online processing involves the direct entry of transactions into the computer system using computer terminals. A significant advantage to this approach is that either the terminal (i.e., a programmable terminal) or the computer can be programmed to edit or verify transactions as they are entered. This allows corrections to be made at a transaction's point of origin. In batch processing, transaction may not reach the computer for editing until hours or days after the transaction has occurred. Consequently, it is more difficult to reconstruct the transaction for error correction.

Another disadvantage of batch processing is that an extra step (i.e., data conversion) is required between the occurrence of the transaction and computer processing. Online processing allows the conversion of the transaction to a machine-readable form as a by-product of conducting the transaction. For example, use of retail or banking terminals to conduct transactions allows the data associated with each transaction to be captured and transmitted to the computer. Figure 4-5*b* illustrates online processing.

In spite of the advantages of online processing, its implementation has been slower than expected. The primary reason for delay has been expense and the need for special-purpose terminals. Fortunately, the cost of computer processing is continually declining. This reduces the overall expense associated with online processing.

The biggest breakthrough for online processing is in the area of distributed processing. Distributed processing involves distributing online processing from large computers to smaller, less costly processing units such as minicomputers and microcomputers.

Minicomputers are now in common use and have proven highly cost effective for online applications. For example, minicomputers can be set up to handle online data entry in each of several department stores. The transaction data associated with each store can be transmitted to the central computer at the end of the day for centralized processing. Note that in this example, transactions are collected *online* for subsequent *batch* processing. Figure 4-6 shows a distributed network that includes four minicomputers.

Computer processing capabilities are also distributed to terminals using microprocessors. Such terminals are said to be *intelligent*, or *programmable.* This technology facilitates the design of industry-oriented devices such as retail terminals and banking terminals. Such devices can perform many functions, independent of a computer. This not only reduces the load on the central computer but also allows the terminal to have stand-alone capabilities which are operational even if the central computer malfunctions. For example, a retail terminal can be used independently to total sales, compute tax, and the like. The terminal can also store sales transaction data on a small cassette tape until the central computer is available.

The reporting of information is greatly enhanced by online processing. Information stored in the computer can be accessed using computer terminals as well as printed on reports. There are two basic levels of online processing—with realtime update or without it.

In a system using online processing without realtime update, transactions are captured online but data files are not updated at the time of capture. Rather, a transaction file is maintained (usually on disk) for subsequent updating of data files. The information displayed on a computer terminal is only as current as the last file update. For example, an accounts receivable department

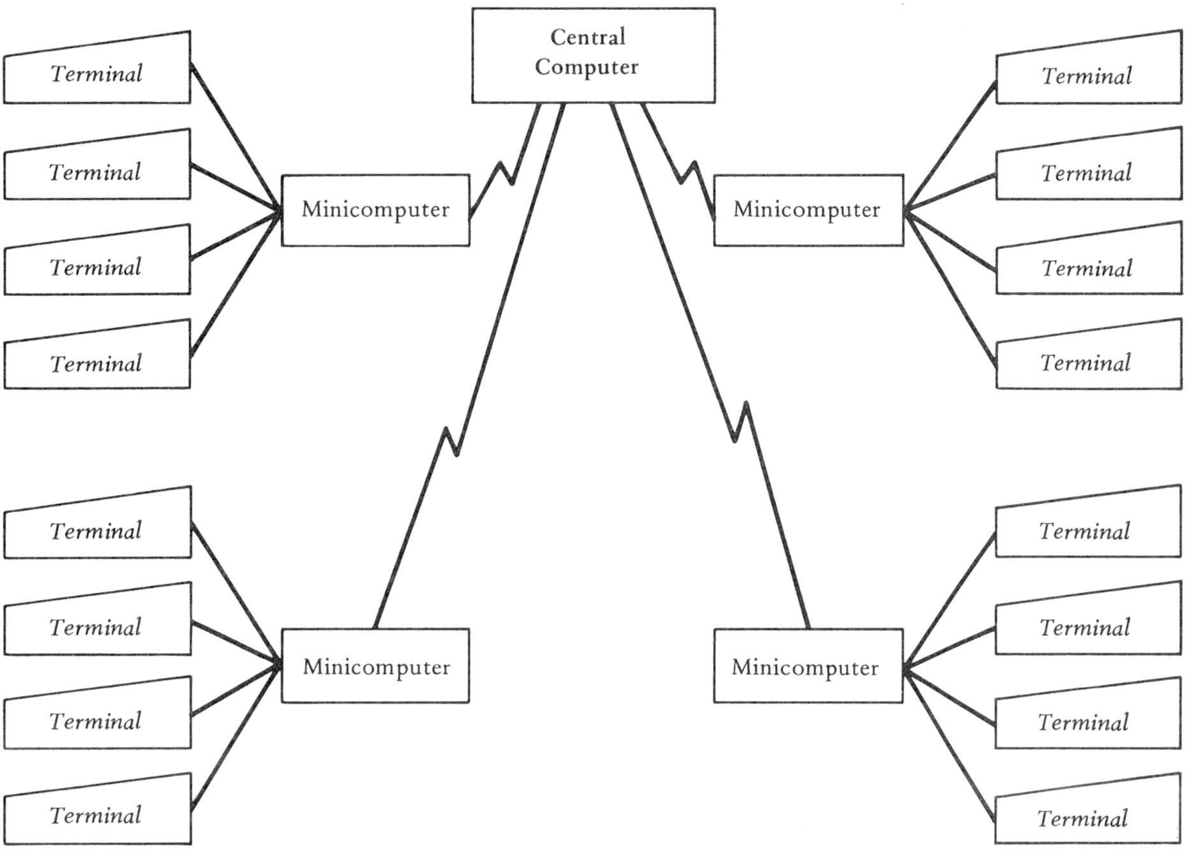

Figure 4-6. *A Distributed Processing Network*

might capture customer charges and payments online but not update the accounts receivable file until the end of the day.

In an online system, data files can be updated immediately to reflect transaction activities. Such online systems are referred to as *realtime* systems. Each input to a realtime system is accepted and processed fast enough to affect subsequent inputs and queries. Because realtime processing is expensive, it should be used only when required. An example of an application requiring realtime processing capabilities is an airline reservation system.

SOFTWARE

In the most general sense, software is computer programs, procedures, and associated documentation used in the operation of computer hardware. The two major categories of software are systems programs and application programs.

Systems Programs

Systems programs are a collection of computer programs used to coordinate and control the overall operation of a computer system. Systems programs perform the following functions:

- Facilitate the loading and execution of application programs.
- Process input and output commands to facilitate interaction with disk drives, tape drives, card readers, printers, and other devices.
- Provide a general library of utility routines for listing the contents of a tape on the printer, copying the contents of a disk to tape, and the like.
- In a multiprogramming system (i.e., where the computer is capable of concurrently running two or more application programs), systems programs insure that each program is serviced by the CPU in a rotating sequence or according to some other priority scheme. They allocate system resources (i.e., primary storage, secondary storage, and input and output devices) among the programs being executed and insure that the programs do not interfere with one another's processing.
- In an online system, the systems programs include a data communications monitor to coordinate and control data transmissions to and from computer terminals.

Application Programs

Application programs are computer programs written for specific applications. These applications range from payroll processing and accounts receivable to forecasting and simulation. Application programs generally depend on operating software during execution. For example, if an application program requires a record from a disk file, the operating software intervenes to retrieve the record from disk and makes it available to the application program.

Originally, most computer users wrote all of their application programs. However, it soon became apparent that some computer applications (e.g., payroll processing) were quite similar at a great number of installations. Rather than "re-invent the wheel," organizations began to buy and sell application programs to organizations with similar requirements. A large number of companies now develop and market specialized application systems for particular industries. It is usually less expensive for an organization to purchase such a system, when available, than to develop its own.

Preprogrammed application "packages" have two major disadvantages. First, they have to be highly generalized to fit a large number of customer requirements. Consequently, they are not as efficient as systems tailored to particular user needs. Second, they seldom exactly fit an organization's requirements. Therefore, modifications are necessary. In some unfortunate cases, the modifications become so extensive that packages have to be discarded.

The emerging trend is clearly in favor of buying more application systems and doing less in-house development. This trend is a result of improvements in the variety and quality of application packages available, combined with the increasing cost and shortage of qualified programmers.

FILES

A file is a collection of related records. For example, a payroll file consists of a payroll record for each employee. The records in a data file are composed of data elements. Some data elements commonly included in a payroll file are employee name, social security number, address, salary, and deductions.

Accessing Files

Files are generally stored on magnetic tape or magnetic disk. Data is accessed using access techniques provided by the operating software. The three basic techniques for accessing files are:

- Sequential access
- Direct access
- Indexed access

The following sections discuss common implementations of these techniques.

Sequential Access

Sequential access pertains to storing and retrieving records in a "one after the other" order. Records are generally stored in ascending or descending order by a record key. A record key is a unique unchanging piece of information such as account number, name, or social security number. Sequential access is the only access technique used with magnetic tape drives, which are, by design, sequential access devices.

The storage and retrieval of records in sequential order is similar to the approach used in manual systems (e.g., of personnel folders stored alphabetically in a filing cabinet). Accordingly, sequential access has traditionally appealed to organizations converting from manual to computer-based systems. This appeal, combined with the early dominance of magnetic tape as a cost-effective storage medium, has led to the use of sequential access files in most initial computing efforts.

Sequential access is used primarily in batch-processing environments. It is particularly effective when transaction activity is evenly distributed among the records in the file. For example, in payroll processing, time cards are recorded for all hourly employees and payroll checks are printed for all employees. When a sequential file is updated, the transactions are sorted into the same sequence as the file and then sequentially matched against the file.

Direct Access

Sequential access is inefficient for batch-processing applications in which only a small proportion of the records in a file are affected by a given batch of transactions. The entire file may have to be passed to update a few records. For online processing, sequential access is inadequate. The time lapse of several minutes generally required to locate a record sequentially is unacceptable.

Sequential access fails to take advantage of two exceptional capabilities of computer technology: speed and direct access. Direct access is an alternative to sequential access that significantly accelerates the process of storing and retrieving records by capitalizing on both the computational speed of the CPU and the access speed of the disk drive. Disk drives are capable of directly accessing any record in a file in a matter of milliseconds. However, to access a particular record, the disk location of the record is required. The location is indicated by the address assigned to it, which is saved when the record is stored or recalculated on the basis of the record key when the record is sought.

In sequential access, the location of a record is a function of its proximity to preceding records. Since processing begins at the first record of the file, it is only necessary to know the address of the location of the first record. All subsequent records follow the first record.

Since direct accessing requires the address of the specific location of a record desired, direct access requires an addressing scheme that computes a unique address for each record. Generally, the record key (e.g., social security number, account number, or part number) must be transformed into a disk storage address.

The most common approach to transforming record keys into storage addresses involves an arithmetic procedure that generates "random addresses" from record keys. There are several randomizing algorithms. The most common algorithm involves the generation of addresses by dividing the record key by a positive prime number (usually, the prime number is the largest prime number that is less than the number of available addresses). The remainder of a division operation is used as the address locator. Figure 4-7 shows the generation of random addresses from record keys when 1000 addresses are available.

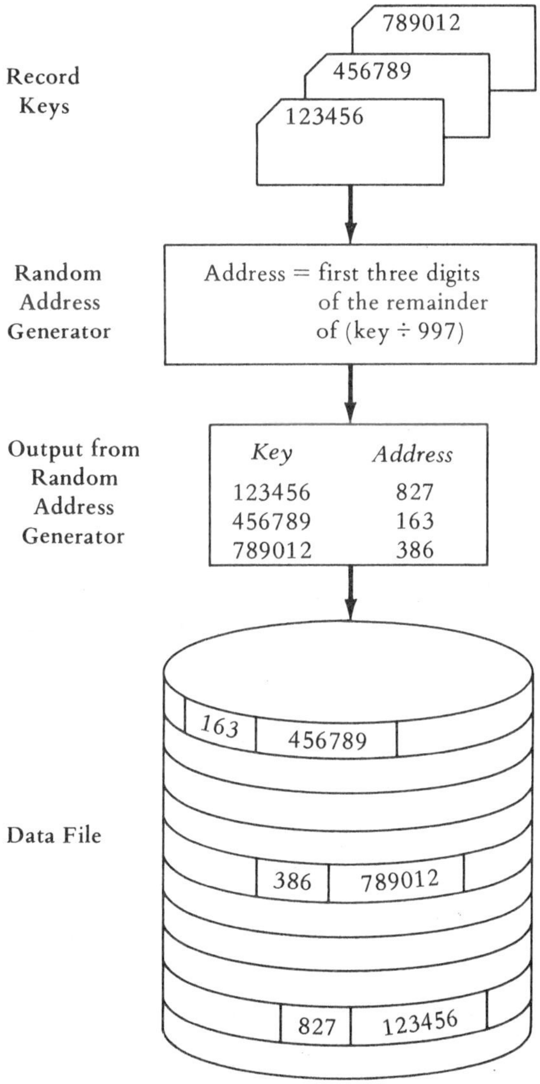

Figure 4-7. *Using Record Keys to Generate Random Addresses on Disk Storage*

A randomizing algorithm always generates the same address for a particular key. Therefore, given the key to a record, the computer can calculate the disk address and then access the record in a matter of milliseconds.

Occasionally, a random address generator generates the same address for two or more keys. The second and succeeding records with duplicate addresses are referred to as *synonyms*. When a synonym occurs, the record having the duplicate address can be stored in a location next to other synonym(s) of the record stored at the computed address, or it can be stored in a general overflow area. In either case, if a desired record is not located at the computed address (this is determined by checking the key stored in the record), a sequential search of synonyms is invoked until the desired record is located. This sequential search slows processing slightly. A good randomizing algorithm generates few synonyms. If a random address generator produces many synonyms for a particular set of keys, analysis and modification of the randomizing algorithm is needed.

Another drawback of a direct-access technique is that, by design, it usually leaves large gaps between records on a disk. This results in wasted disk space. Some of the gaps may be consumed by synonyms, but considerable wasted space still remains. An offsetting advantage is the incredible speed with which records can be accessed, regardless of the size of the file.

Perhaps the most challenging concept to grasp with respect to direct access is that the physical locations of records bear no relationship to the logical view of the data. The random generation of addresses physically scatters records throughout the disk such that, without knowledge of the randomizing algorithm used to transform the keys, locating a record requires a sequential search through a non-sequentially-ordered file. The use of a randomizing algorithm is an extraordinary deviation from the way that files are maintained manually. However, it is a highly suitable technique for computer-processed files. Any one record of several thousand or even several million can be located instantaneously.

Indexed Access

Indexed access pertains to using tables of key fields that provide the disk addresses of records stored in a file. Indexed-access techniques are used with either sequential or direct-access files.

The simplest form of indexed access is referred to as indexed sequential. This technique involves the use of indexes to segment a sequentially stored file to facilitate quicker access. To access a specific record, a search is made of the index file to determine the address of the first record of the segment in which the record is located. A direct access is then made to that address. This positions the disk's read/write head in the proximity of the record desired. A sequential search of the file segment is then invoked until the record is located.

Since the index file is much smaller than the actual data file, indexed-sequential processing greatly accelerates the locating of a specific record; substantial sequential processing is eliminated. For large files, multiple levels of indexes (i.e., hierarchical indexes) are used to more quickly locate records. Figure 4-8 shows a hierarchical indexing scheme using three levels. The major-level index points to the intermediate-level index, which points to the minor-level index, which points to the actual data file.

Indexed-access techniques can also be used to access records by means other than record keys. This provides flexibility that is especially useful for querying a data file. For example, the personnel office can request the records of all employees who are

Figure 4-8. *A Hierarchical Indexed Sequential Structure*

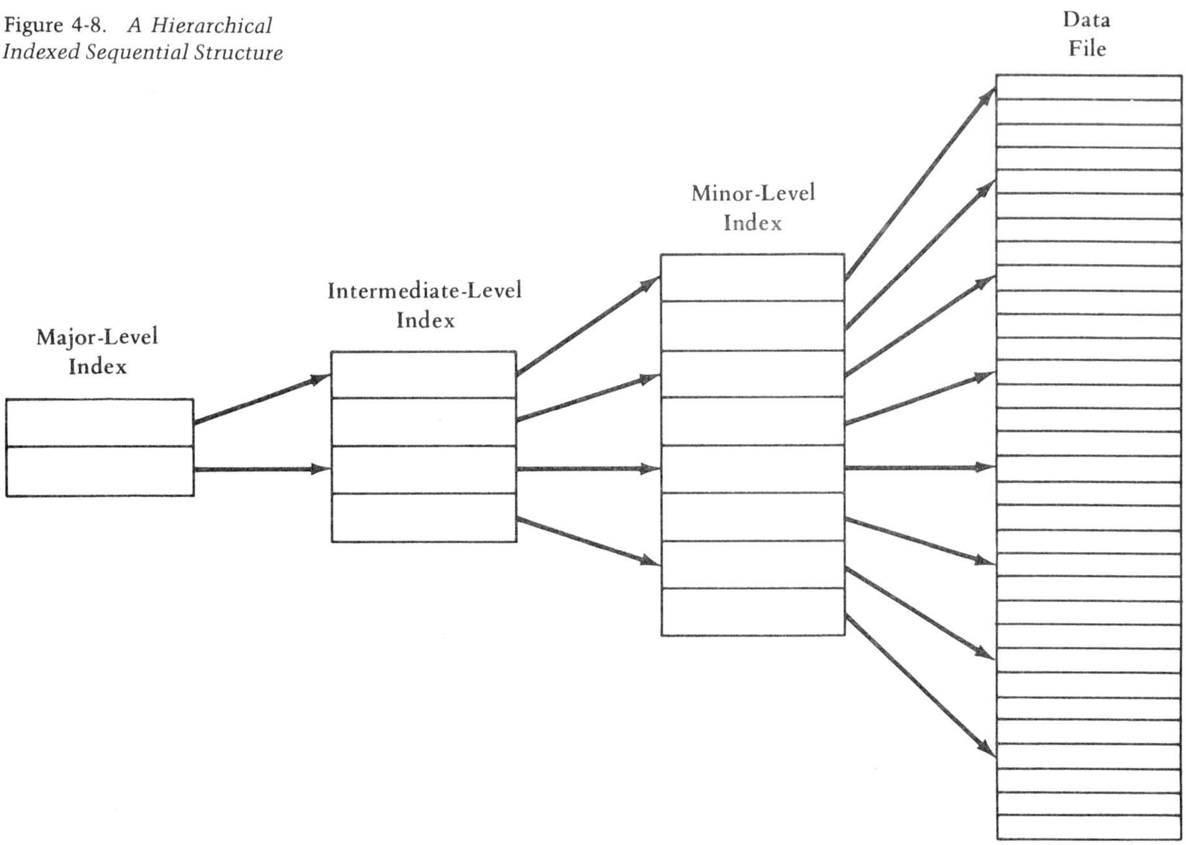

accountants. Such access to records is accomplished by setting up indexes based on one or more fields of the records. In the same way that the index in this book points to each page that contains a particular word, a field index points to each record that contains a certain field value. Figure 4-9 illustrates a job classification index for a personnel file.

Indexes can be set up for several record fields. This allows even more discrimination in making a data-file query. Instead of just requesting the records of all employees who are accountants, the query can also specify that the employees have a college degree and two years of experience within the organization. One way of responding to this query requires that an index exist for job classification, education, and number of years within the organization. The query is satisfied by identifying all records pointed to by the first index, and then progressively eliminating any records that are not pointed to by the second and then the third index. After this process of elimination, only records that satisfy all the selection criteria remain. These records can then be directly accessed using the addresses provided by the indexes.

The use of indexed access on fields other than the record key is more suitable for higher-level decision-making and planning functions than for routine transaction processing. It allows convenient and less restrictive retrieval of information, but considerable programming time and computer processing are required to set up and maintain the indexes as records are added to and deleted from the data file.

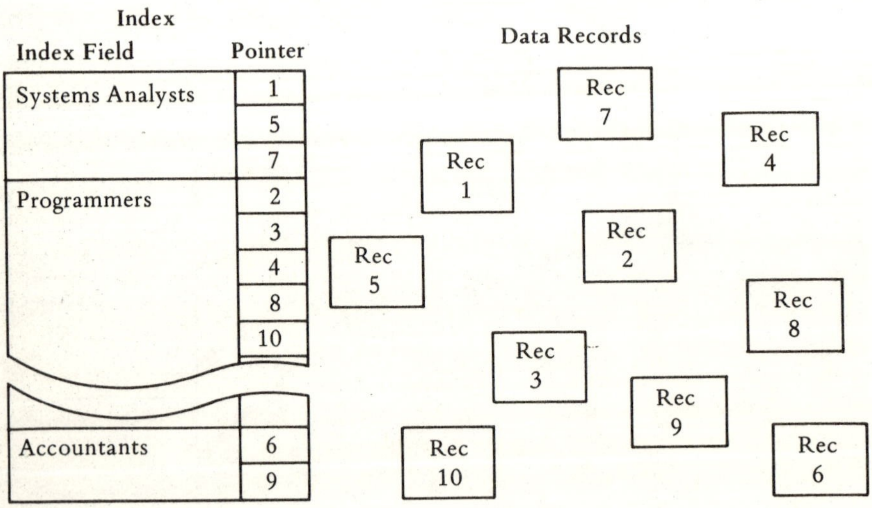

Figure 4-9. *Job Classification Index for Accessing Employee Data Records in a Randomly Organized File*

Database Management Systems

The preceding discussion of access techniques was oriented toward a single-file environment. At the end of Chapter 3, the concept of a management information system (MIS) was introduced. In that discussion, an MIS was portrayed as a federation of information systems, integrated as necessary. Database technology is a step beyond file-access techniques and is a key factor in achieving an MIS. It provides software to facilitate the integration of the various data files of an organization.

Database technology is still in a somewhat formative stage. Not all data processing installations have implemented or are implementing it. But database software offers considerable potential. The use of database technology is clearly an emerging trend in information systems. Database technology is covered in the third section of the text, "Advanced Systems Analysis."

PROCEDURES

Procedures are specific instructions for both the development and the operation of information systems. The accepted means of documenting and conveying procedures is through procedures manuals. Such manuals can be viewed as "cookbooks" that carefully lead personnel through the necessary and accepted steps to complete a task or resolve a problem.

Systems Development Procedures

Systems development procedures provide standards for systems analysis, systems design, programming, and documentation. Adherence to development standards insures uniformity of information systems. This uniformity is important because most information systems take over a year to develop and then have a useful operating life of several years. The original systems analysts and programmers frequently change responsibilities or organizations during either the development or the operation of a system. When this happens, new personnel are assigned to take over their responsibilities. The uniformity established through development procedures greatly eases transition of responsibilities.

Some organizations have experienced considerable hardship because of the loss of key systems analysts and/or programmers when systems development procedures did not exist or were not enforced. For example, assume a payroll system must be modified to implement new tax tables. Without procedures to explain how or which computer program computes taxes, this modification is difficult. Payroll processing may be delayed.

Operating Procedures

Operating procedures help to insure that the processing of data is accomplished in an efficient, effective manner. These procedures comprise specific instructions for data collection, preparation, and processing, and for report distribution. For example, organizational departments are told how to collect and prepare payroll transactions; data-entry operators are provided card-punching instructions; and computer operators are told which computer programs and files to use.

Operating procedures are also used for control purposes. They document the actions to be taken when errors are discovered in the collection, preparation, processing, and distribution activities. For example, a payroll clerk can refer to a procedures manual to determine what action to take if payroll totals are out of balance.

PERSONNEL

The personnel in a data processing or information systems department can be broadly categorized as either development personnel or production personnel. The development personnel include systems analysts and programmers involved with the analyzing, designing, implementing, and evaluating of information systems. Once a system has been developed and is operational, it is released to production personnel for ongoing scheduling and production processing. Production personnel include computer operators, data preparation/entry operators, and production control personnel.

Position Descriptions

Table 4-1 provides basic job descriptions for both development and production personnel. Within each job category, varying degrees of experience are generally recognized by designating positions at junior, senior, and lead levels.

The increasing complexity associated with computer-based information systems has resulted in the need for further specialization of the basic data processing positions and for entirely new positions. Tables 4-2 and 4-3 depict the further specialization and the creation of new data processing positions.

Table 4-1. *Position Descriptions for Basic Data Processing or Information Systems Personnel*

Development Personnel	
Systems Analyst	Analyzes, designs, implements, and evaluates computer-based processing systems to process transactions and provide information to meet organizational requirements.
Programmer	Designs, codes, tests, and debugs computer programs consistent with systems design specifications.
Production Personnel	
Computer Operator	Monitors and controls the operation of computer equipment.
Data Preparation/Entry Operator	Operates card-punch/key-to-disk/key-to-tape equipment to prepare data in machine-readable form and/or operates computer terminals to directly enter data into a computer system.
Production Controller	Records and maintains control information on the receipt and processing of transactions, and the subsequent distribution of generated reports.

Organization and Management of Data Processing Personnel

Data processing departments are organized in a variety of ways. Figure 4-10 illustrates a common approach. With this approach,

Table 4-2. *Further Specialization of Basic Data Processing Positions*

Basic Position	Specialized Position	Description
Systems Analyst	Information Analyst	Specializes in defining information requirements. Has specialized knowledge of organizations, management, and decision making.
	Systems Designer	Specializes in the hardware and software necessary to implement computer-based information systems. Has stronger technical orientation than information analyst.
	Production Analyst	Specializes in "trouble-shooting" and coordinating systems that are in production status. Is generally responsible for several production systems and works with user departments on issues of scheduling, training, and systems maintenance and modification.
Programmer	Application Programmer	Specializes in designing, coding, testing, and debugging new computer programs.
	Maintenance Programmer	Specializes in the maintenance and trouble-shooting of existing computer programs.
	Systems Programmer	Specializes in operating software. Has high level of technical proficiency.

Table 4-3. *New Data Processing Positions*

Position	Description
Database Administrator	Administers and controls organizational databases.
Librarian	Maintains and controls systems and program documentation. Controls use and insures backup of files stored on disks and tapes.

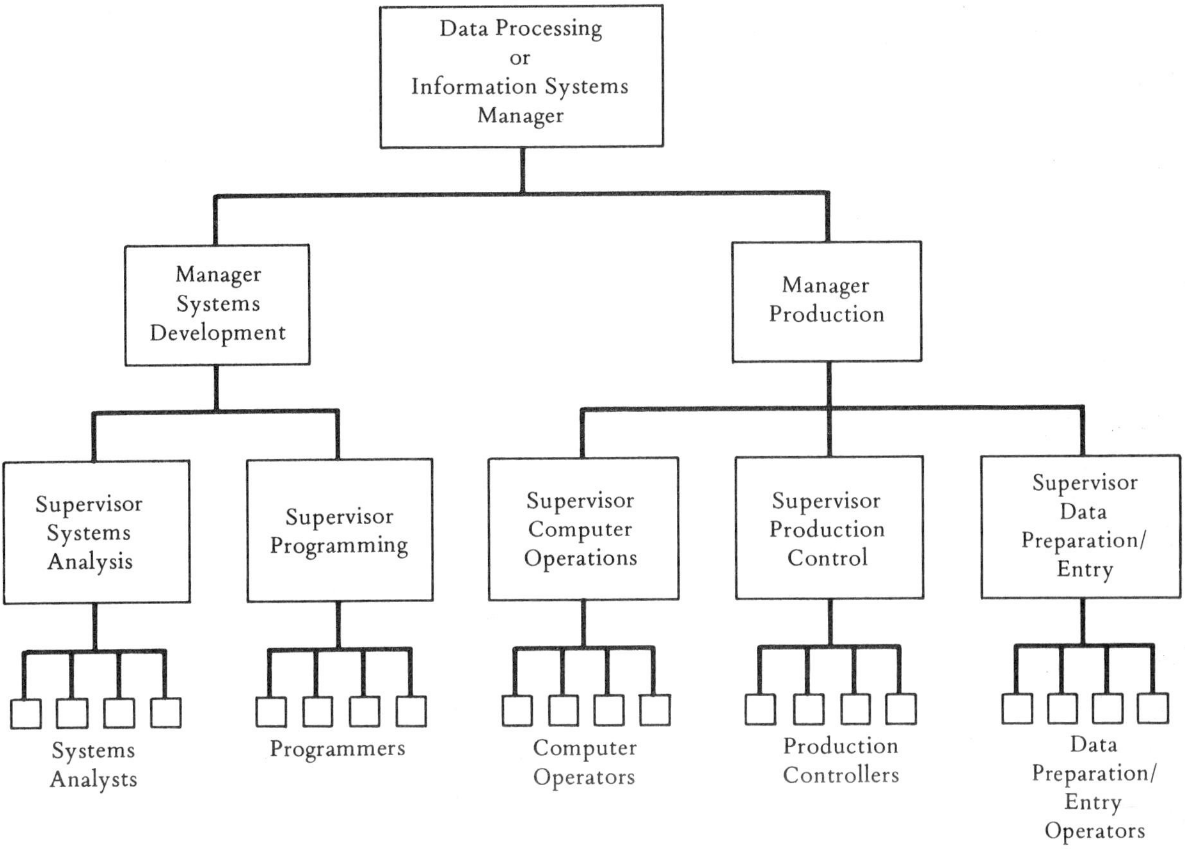

Figure 4-10. *A Data Processing or Information Systems Department*

the production and development functions are separated into departmental sections.

The major variation of the structural organization depicted in Figure 4-10 involves having programmers report directly to systems analysts rather than pooling the programmers. This variation encourages the programmers to identify with projects they are working on. However, it inhibits flexibility: the shifting of programmers to projects assigned to different systems analysts cannot be accomplished without organizational change.

The segregation of the development and production functions is beneficial from the contingency management perspective discussed in Chapter 3. The production functions of computer operations, data preparation/entry, and production control are relatively routine and predictable in operation. Work requirements are reasonably structured and defined. Employees should punctually follow established instructions and procedures. Therefore, the production function lends itself to management in a classic orientation for what is a relatively closed/stable/mechanistic environment.

Conversely, the development functions of systems analysis, design, and programming are quite dynamic. User requirements are difficult to precisely define and difficult to translate to computer technology. The development department lends itself to a human relations orientation for what is an open/adaptive/organic environment. Accordingly, systems analysts and programmers are generally allowed more latitude in their decision making, working hours, professional development, and so forth, than is feasible for production personnel.

Data processing or information systems departments report to different places in organizations. Traditionally, they have reported to the chief financial officer, primarily because financial applications have been the first to be computerized in most organizations. As new applications are approached, other departments become involved with computer-based information systems. Data processing or information systems functions appear to be evolving into independent departments reporting directly to chief executive officers.

SUMMARY

The major subsystems of an information system consist of hardware, software, data files, procedures, and personnel. These subsystems interact to transform data into information suitable for decision making.

The basic hardware components are input/output devices, secondary storage devices, and the central processing unit. The major forms of processing are batch processing and online processing. Distributed processing is an emerging architectural concept for structuring data processing hardware. Distributed processing involves distributing processing capabilities to minicomputers and microprocessors.

Software is categorized as either operating software or application programs. Operating software coordinates and controls the overall operation of the computer system. Application programs are written to process specific applications.

Files are collections of related records. The records within data files can be accessed using sequential, direct, or indexed access. A data base system is a collection of files and software organized to facilitate the interrelating of the files of an organization.

Procedures are specific instructions for both the development and operation of systems. Procedures manuals are set up to instruct personnel on how to complete a task or resolve a problem.

The personnel in a data processing or information systems department can be categorized as either development or production personnel. Development personnel primarily consist of systems analysts and programmers. Production personnel primarily consist of computer operators, data preparation/entry operators, and production controllers.

EXERCISES

1. Identify and define the major hardware components in a computer system.
2. Discuss the major advantages and disadvantages of batch and online processing.
3. What technological developments have enhanced the viability of online processing?
4. Discuss the difference between systems programs and application programs.
5. Why is application software increasingly being purchased rather than developed in-house?
6. Define and discuss the relationships among data elements, records, files, and data bases.
7. Define and discuss the three major types of file access. Which access techniques are suitable for online processing?
8. Discuss the role of procedures in an information system.
9. Discuss alternative ways of organizing systems analysts and programmers.

SELECTED REFERENCES

Bateman, Barry L., and James C. Wetherbe. "Production Analyst: A New Position for an Old Problem," *Data Management,* Dec 1974, pp. 30–33.

Date, C. J. *An Introduction to Database Systems.* Reading, Mass.: Addison-Wesley, 1975.

Davis, Gordon B. *Management Information Systems: Conceptual Foundations, Structure, and Development.* New York: McGraw-Hill, 1974.

"Distributed Computing: A Growing Concept," *Infosystems* Vol. 22, No. 8 (Aug 1975), pp. 32–34.

"Distributed Intelligence in Data Communications," *EDP Analyzer* Vol. 11, No. 2 (Feb 1973).

Ferrara, Raymond, and Richard L. Nolan. "New Look at Computer Data Entry," *Journal of Systems Management,* Feb 1973, pp. 24–33.

IBM. *IBM System/370 System Summary.* GA22-7001-6, File No. S/370-00.

"In Your Future: Distributed Systems?" *EDP Analyzer* Vol. 11, No. 8 (Aug 1973).

Kelly, Neil. "Bank of America Goes Distributive," *Infosystems,* Mar 1977, pp. 46–48.

NCR. *File Management Software Summary.* Dayton, Ohio: The National Cash Register Company, 1972.

Nolan, Richard L. "Computer Data Bases: The Future Is Now," *Harvard Business Review,* Sept–Oct 1973, pp. 98–114.

Reside, Kenneth D., and Theodore J. Seiter. "The Evolution of an Integrated Data Base," *Datamation,* Sept 1974, pp. 57–60.

Ritchie, Robert O. "Intelligent Terminals and Distributed Processing," *Computer Decisions,* Feb 1975, pp. 36–40.

"Salary Survey 1973," *Infosystems* Vol. 20, No. 9 (Sept 1973), pp. 34–49.

Schubert, Richard F. "Directions in Data Base Management Technology," *Datamation,* Scpt 1974, pp. 48–51.

Wetherbe, James C., and C. J. Whitehead. "A Contingency View of Managing the Data Processing Organization," *MIS Quarterly,* Mar 1977, pp. 19–25.

White, Robert B. "A Case against Large-Scale Computers," *Infosystems,* Aug 1975, p. 34.

SECTION II The Processes and Techniques of Systems Analysis

The material covered in the first section of this text has provided the conceptual foundation for a more pragmatic view of the processes and techniques of systems analysis discussed in this section of the text.

Chapter 1 provided an overview of the systems development cycle (see Figure 1-1). The eight steps involved in the systems development process are allocated among the five chapters in this section as shown below.

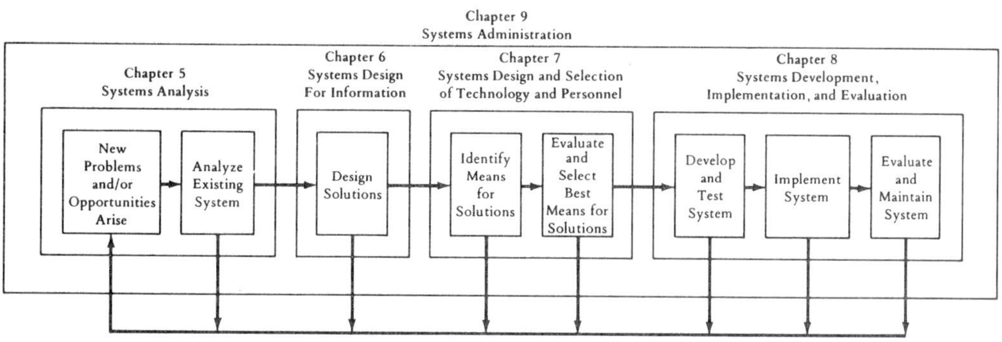

Chapters 5 through 8 cover the material contained within their respective boxes. Chapter 9 covers topics related to the overall administration of the systems development process. A case study centered around payroll processing follows Chapters 5 and 6. It applies the systems design concepts and techniques introduced in these chapters.

5 Systems Analysis

This chapter discusses the "analysis" dimension of systems analysis. The particular areas of interest are related to the methodological investigation of a problem or opportunity. The problem or opportunity is identified and separated into related components for further detailed study.

IDENTIFICATION AND DEFINITION OF PROBLEMS AND OPPORTUNITIES

Before a problem or an opportunity can be analyzed, it must be clearly identified and defined. The first section of this chapter presents concepts pertinent to identifying and defining problems and opportunities.

Reaction and Opportunistic Surveillance

There are two basic approaches to management decision making that are relevant to defining problems and opportunities. These two approaches are called *reaction* and *opportunistic surveillance.*

Reaction involves addressing problems as they arise. This is a "management by exception" approach. It is commonly employed. For example, if customers begin to complain about the tardiness of order processing, management reacts by attempting to resolve the problem. Otherwise, management "leaves well enough alone."

Opportunistic surveillance involves continually seeking opportunities that might be beneficial to the organization. Rather than leave well enough alone, management seeks continual improvement. This is a much more aggressive posture, and it can pay considerable dividends. For example, management may see an opportunity to acquire more customers by implementing a new order processing system that will provide customer deliveries superior to those provided by competitors.

Generally, the systems analyst is not called in on a major problem or opportunity until after the problem or opportunity has been identified by management. However, once the systems analyst is involved, he or she should maintain an opportunistic surveillance as well as a reactionary orientation. New opportunities and problems continually arise in organizations. Many of them are subsets of an original opportunity or problem identified by management. The systems analyst should be alert to such developments and call them to management's attention so that they can be properly considered and addressed.

Object System and Information System

A distinction that is sometimes subtle but nevertheless important is the distinction between an object system and an information system. An *object system* is a physical process whose purpose is to achieve one or more organizational goals. Examples of object systems include manufacturing of products and selling of merchandise.

An *information system* is a physical process that supports an object system in achieving organizational goals. The information system runs parallel to the object system to provide both documentation and information for decision making pertinent to the management and operation of the object system. For example, in a manufacturing plant, work orders are used to document the work flow, and production schedules are used to plan the work flow. In a retail store, sales are documented by being entered into cash registers or retail terminals, and sales information is used for merchandising decisions.

A helpful and important way to differentiate between object systems and information systems is to keep in mind that without an object system there is no need for an information system. The purpose of an information system is to facilitate the operation of an object system. Indeed, the cost of operating the information system should be offset by the savings realized because the object system can be (and is) more efficiently and effectively managed.

Generally, the systems analyst accepts the object system as given. However, there are occasions in systems analysis when potential improvements in the object system are apparent to the systems analyst. For example, rearranging work stations in a manufacturing plant might save considerable time in transporting goods-in-process through work stations. In such cases, the systems analyst should consider analyzing and designing improved methods for the operation of the object system as well as the information system.

Identifying Problems and Opportunities

There are a multiplicity of problems that can occur and opportunities that can arise in organizations. Management generally expresses problems and opportunities in terms of an "end result," or "bottom line." These end-result problems and opportunities can generally be categorized as discussed in the following paragraphs. Though these categories are not exhaustive, they provide a basic framework for use in identifying problems and opportunities. The categories are:

Performance
Information
Economy
Control
Efficiency
Security

Note that the first letters of the names of these categories can be combined to form the acronym **PIECES.** Indeed, each category can be viewed as a piece of the problem or opportunity. The categories occasionally overlap, but each is sufficiently different to warrant a separate discussion. The categories also interact. Consequently, a particular problem or opportunity may belong within one category or span several categories.

In the following discussion, problems and opportunities related to both object systems and information systems are discussed.

Performance

Performance problems and opportunities pertain to the effectiveness of the operation of a process. Primary performance criteria are throughput and response time.

Throughput is the total volume of work performed over a given period of time, for example, the number of orders processed per day by the order processing department. The *response time* of a system is the average time between submission of a request and return of the result. In an online system, response time is the time between the end of a block of user input and the display of the first character of the computer's response at the terminal.

Throughput and response-time dimensions should be both separately and collectively considered and adjusted to achieve a desired performance level. For example, a bank may service a large number of customers (good throughput) but many customers may endure considerable delay before being serviced (poor response time). In any given situation, both throughput and response time can be poor, or one can be good while the other is poor. In either case, the result is a performance problem.

Even if existing performance is satisfactory, opportunities to improve performance may arise. Through such improvements, an organization may gain an important advantage over its competitors. For example, consider the first banks that installed electronic tellers for after-hours check cashing. In general, the management of these banks took an opportunistic surveillance approach to improve service beyond existing levels.

Information

Information is used in decision making. Therefore, information problems and opportunities are generally related to improving the relevance, form, timeliness, or accessibility of information. To be useful in decision making, information must be *relevant* to the decision being made. Information that is not relevant is unnecessary; it can be a deterent to decision making. The irrelevant information has to be sifted through to locate the relevant information.

The more closely the *form* of relevant information matches the requirements of the decision maker, the more effective the information. The language, format, accuracy, and presentation of the information are key variables in providing information that is comprehensible and usable to the decision maker.

To be used for decision making, information must be available prior to the decision-making process. Therefore, information should be available in advance of the *time* of need. Otherwise, the decision is delayed or made without the information. Besides being timely, information must be *accessible* to the decision maker. One of the significant contributions of online systems is the improvement of both the timeliness and the accessibility of information.

Economy

Economic problems and opportunities pertain to the costs of a process. The two key variables to be balanced when considering economy are service level and excess capacity. The more excess capacity, or slack resources, available for a process, the more quickly responses can be made to demands for goods or services. However, higher service levels are achieved with losses in economy because excess capacity is costly.

The proper balance between service level and excess capacity is contingent upon the organizational process. Generally, the more unpredictable the demands for service are, the greater the excess capacity required to insure good service. For example, ambulance services and fire departments have considerable excess capacity so that they can quickly respond to demands for service. Conversely, an assembly line in a manufacturing plant attempts to have minimal excess capacity to keep down costs.

Though the appropriate level of excess capacity should be considered on an individual application basis, one guideline is reasonably universal: Excess capacity should be reduced to the point that it does not adversely affect the desired service level.

Control

Control is used to monitor or regulate processes. Control is achieved by checking for deviations from planned performance. When deviations occur, some form of control mechanism is required to invoke corrective action. For example, a manager responsible for control of a process is primarily concerned with deviations outside of allowable limits. Therefore, the manager's information system should provide exception reporting on problems. This reduces the information-processing requirements of the manager and allows the manager to concentrate on items requiring corrective action or further investigation.

Control can present problems or opportunities in one of two ways—lack of control or too much control. Lack of control can result in a system that is out of control. For example, a key factor in the design of an accounting system is the concept of auditing—the formal, periodic examination and checking of financial records to verify their correctness. Without auditing procedures, many opportunities exist for fraud and embezzlement. Fraud and embezzlement have been facilitated somewhat by computer technology due to some auditors' lack of computer expertise. This lack of expertise impedes their ability to audit properly that portion of the accounting system performed by the computer. Recently, many auditors have expanded their auditing skills to include an understanding of computer-based accounting records. By doing so, they have put "teeth" back into the auditing process.

Too much control can result in diminishing returns. When an organization spends more on control than it can derive as benefits from the control, the control is no longer cost effective. Also, excessive control can become "red tape," negatively impacting productivity. Consider a university registration system that requires each student to get his or her class schedule approved by an advisor, department chairperson, and dean. Few deans or department heads have the time to review fully each student's schedule before signing it. Consequently, requiring additional signatures adds little if any control, though it adds inconvenience to the registration process.

When control is lacking, problems can be resolved or opportunities gained by increasing control. When control is excessive and unwarranted, reducing control can reduce costs and inconvenience.

Efficiency

Efficiency refers to producing a product or providing a service with minimum waste. It can be expressed as a ratio of effective work to energy expended in working. More simply, efficiency is output divided by input.

Understanding the significance of the efficiency ratio is a prerequisite to truly appreciating the meaning of efficiency. Efficiency is often confused with economy. *Economy* refers to the amount of resources committed to a process; *efficiency* refers to minimizing waste in the use of those resources.

For example, an organization might install slower, less expensive terminals to economize when setting up an online system. Given that the organization has slow-speed terminals, the organization should insure that the terminals are used efficiently by locating them where they are most needed, insuring proper training and productivity of terminal operators, and keeping the terminals in good working condition.

It is important in systems analysis to recognize the roles of economy and efficiency in the amount of output produced. For example, if management is concerned with increasing output, management must increase production resources (e.g., manpower and equipment) *and/or* increase the efficiency of the use of the resources. If the process is already efficient, additional output requires additional resources. If the process is inefficient, efforts to improve efficiency should be pursued before committing additional resources.

If management attempts to reduce costs by reducing resources, then either it must settle for less output, or efficiency must be increased to offset the loss of resources. If the process is already efficient, reduced output is inevitable when resources are reduced.

Security

Security pertains to the safety or protection of resources. In an information system, security measures are concerned with safeguarding the following resources:

- Information
- Software
- Documentation
- Equipment
- Personnel
- Facilities

These resources must be protected from natural disasters, man-made disasters, theft, accidents, and misuse.

Good security measures are designed to *avoid* problems. However, the types of problems that security measures attempt to avoid seldom occur in organizations. Consequently, there is a tendency for organizations to become lax in maintaining and enforcing security, thus leaving themselves vulnerable. To avoid vulnerability, most organizations employ internal and/or external auditors to review both the viability of security procedures and their enforcement.

Security procedures generally consist of both *controlling* and *documenting* the access to and utilization of information system resources. Doing so deters security violations. Documenting also provides records that can be used for investigating and isolating security violations if they occur. For example, in an online system, a transaction log can be kept of all file changes. The transaction log documents each file change, when it occurred, the terminal used, and the terminal operator making the change. Consequently, any tampering with files can be accurately isolated as to who, what, when, and where.

Since security measures cannot prevent all problems, they must also address recovery procedures. Recovery is generally provided for by maintaining backup copies of files, software, and documentation at a site remote from the computer center. Also, agreements are usually made with other computer centers to use their computers if necessary. If an organization's computer, files, documentation, and software are totally destroyed by a disaster such as a fire, the organization can continue processing by using its backup files, software, and documentation on another organization's computer.

Defining Problems and Opportunities

It is not usually difficult to identify general problem areas or possible opportunities. Since perfection is rare, there is seldom an organizational process that is beyond improvement in some aspect of its performance, information, economy, control, efficiency, or security (PIECES).

It is generally more difficult to define with specificity the different dimensions or the various cause and effect relationships that are creating problems or preventing opportunities. For instance, late processing of customer orders may be identified as a problem. On the surface, this may appear to be a performance problem related to response time. However, there are many potential causes of the problem. They include:

1. Salespersons failing to promptly turn in customer orders.
2. Incompetent management of the order processing department.
3. Keypunch operators giving too low a priority to the orders (e.g., they may give higher priority to payroll transactions).
4. A morale problem in the order processing department.
5. An insufficient number of keypunch operators.
6. The computer programs that process orders have several "bugs" in them, resulting in some orders being processed incorrectly.
7. Computer capacity is insufficient to handle the workload.
8. Due to poor routing, information on inventory fluctuations is not timely enough to prevent frequent stockouts.
9. The inventory manager is not aware of late deliveries because this information is not reported.
10. The credit department does not review customer credit checks until just before orders are to be shipped. Therefore, an order has to be held if any questions on credit arise.

The preceding list indicates that to fully ascertain the cause of a problem or the prevention of an opportunity requires a thorough understanding of both the object system and the information system.

Magnitude of Problem or Opportunity

Problems and opportunities vary in complexity. In many cases, a problem or an opportunity can be addressed with minimal difficulty and time. This usually occurs in situations where the existing system is well understood, and new problems or opportunities are rather glaring. Consequently, the appropriate solution is apparent. For example, if late processing of customer orders is due to new salespersons failing to promptly turn in orders, this problem should be easy to resolve. The systems analyst may simply write a memo to the sales manager explaining to him or her the importance of promptly turning in orders and requesting that appropriate action be taken (see Figure 5-1).

MEMORANDUM

December 18, 1979

TO: Mr. Green
Sales Manager

FROM: John Goode
Systems Analyst

SUBJECT: Delayed Processing of Customer Orders

During the past few months, Mr. Hickson (the company's president) has received numerous complaints about the tardiness of order processing. He requested that I review the work flow of the order processing system to isolate the cause of these delays.

The major bottleneck appears to be a delay in salespersons' turning in of customer orders. In some cases, orders are not being turned in until a week after the orders are requested by customers. Most of these late orders appear to be caused by recently employed salespersons who are possibly not aware of the importance of turning in all orders promptly.

Continued delays in order processing can adversely affect customer relations and future sales. Your cooperation in resolving this problem will be greatly appreciated. If I can be of any assistance, please let me know.

cc: Mr. John Hickson, President
Mr. Joe Bond, MIS Manager

Figure 5-1. *Use of Memo to Resolve Information System Problem*

For more complex problems and opportunities, a more comprehensive analysis is required. For example, if the information system is not providing needed information and/or the technology used in the information system is obsolete, problem resolution requires considerably more effort. In such cases, problem resolution or opportunity achievement requires a major overhaul of an existing system or the development of a new one.

Feasibility Study

When complex problems and opportunities are to be defined, it is generally desirable to conduct a preliminary investigation or a feasibility study. A feasibility study is conducted to obtain an overview of the problem and to roughly assess if feasible solutions exist prior to committing substantial resources to a project. During a feasibility study, the systems analyst usually works with representatives from the department(s) expected to benefit from the solution.

The primary objective of a feasibility study is to assess three types of feasibility.

1. **Technical feasibility.** Can a solution be supported with existing technology?
2. **Economic feasibility.** Is existing technology cost effective (i.e., will the cost be offset by the benefit(s))?
3. **Operational feasibility.** Will the solution work in the organization if implemented?

By intent, the feasibility study is a very rough analysis of the viability of a project. It is, however, a highly desirable checkpoint that should be completed before committing more resources. The feasibility study answers a basic question: Is it *realistic* to address the problem or opportunity under consideration?

The final product of a successful feasibility study is a project proposal for management. The contents of this report may include, but are not restricted to, the following items:

1. Project Name
2. Problem or Opportunity Definition
3. Project Description
4. Expected Benefits
5. Consequences of Rejection
6. Resource Requirements
7. Alternatives
8. Other Considerations
9. Authorization

Project Team

If the project proposal is approved, management generally organizes a project team and assigns responsibility for the project to it. The specific personnel or type of personnel to be included on the project team should be defined under the resource requirements section of the project proposal.

The organization of a project team is a prerequisite to good systems analysis. Different types of expertise are required to thoroughly consider all system variables. For example, the system analyst provides primarily computer technology-oriented skills. Without adequate representation from the various departments affected by the information system, various dimensions of their needs and requirements are apt to be overlooked.

Project team participants should be given sufficient release time from their normal duties to allow them to contribute the necessary level of participation to the project. However, the release time should be provided only for the duration of the project. Upon project completion, the participants should return to their normal duties.

An interesting debate about project teams pertains to who should be the project leader. Some managers contend that a systems analyst should provide the leadership; others contend that this role is better filled by a management-level representative from the department(s) that will use the system. Projects tend to be completed faster when a systems analyst is in charge. However, users are generally more committed to and more satisfied with the results of projects when they are directly responsible. When a systems analyst is in charge, users have a tendency to give him or her too much latitude in designing the system (often because users do not understand much of the technology) and then decide whether or not they like the completed system. Unfortunately, the systems analyst may design a system that works right, but is not the right system (i.e., it does not adequately satisfy the user's information requirements). When users are directly responsible for a project, they are inherently more involved in the design of the system. Their involvement improves the probability that the right system is designed.

ANALYSIS OF EXISTING SYSTEM

The identification and definition of a problem or an opportunity, and the organization of a project team, set the stage for a detailed analysis of the existing system. Concepts of analysis are presented in this part of the chapter.

Fundamental Activities for Analyzing Systems

Analysis is the process of separating a whole into its parts to allow examination of the parts. The examination leads to an understanding of their nature, function, and interrelationships. Considerable effort is required to conduct an analysis of an information system. The overall system must be defined. Then it must be separated into its subsystems or processes for further analysis. For example, an order processing system may be separated into subsystems of order collection, entry, editing, posting, shipping, and reporting.

The four basic activities used to gather information about an information system are listed in Table 5-1 and discussed further below.

Documentation and Observation

A review of available documentation is a logical starting point when seeking insight into a system. A documentation review allows project team members involved in systems analysis to gain

Table 5-1. *Information-Gathering Activities*

Activity	*Explanation*
Review of Documentation	• Reviewing recorded specifications that describe the objectives, procedures, reports produced, equipment used, etc., in an information system.
Observation	• Watching the object system and/or the information system in process to note and record facts and events about their operation.
Interviews	• Meeting with individuals or groups to ask questions about their roles in and their use of an information system.
Questionnaires	• Submitting questions in printed form to individuals to gather information on their roles in and use of an information system.

some knowledge of a system before they impose upon other people's time. Unfortunately, documentation seldom completely describes a system, and it often is not up-to-date. The current operation of the system may differ significantly from what is described. Therefore, after a review of available documentation, a logical next step is to observe the operation of the system. Observation provides a more tangible perspective of what is described in the documentation. It also brings to light aspects of the documentation that are incomplete and/or outdated.

Interviews and Questionnaires

Having gained as much knowledge as is reasonably possible without imposing on other personnel, those involved in systems analysis can intelligently use interviews and/or questionnaires to gather additional information. Techniques of interviewing and questionnaire design are topics for extensive study. However, a few key concepts should bc noted.

Interviews are generally preferable to questionnaires for information gathering because they allow thorough questioning and discussion. In an interview, not all questions have to be determined in advance (as they do in a questionnaire). However, when information has to be gathered from a large number of people and tabulated, questionnaires are usually preferable for efficiency reasons.

The two basic formats for questioning are categorized as open ended and close ended. The difference between the two types of questioning is like the difference between essay and multiple-choice exams.

Open-ended questions allow latitude to the persons responding. Consequently, information gathered from open-ended questions may be creative and rich in content. Examples of open-ended questions are:

1. Are there areas of dissatisfaction with the existing information system? If so, what are they?
2. Do you have suggestions for improvements if a new system were designed? If so, list your suggestions.

Close-ended questions are more restrictive in the latitude allowed to a respondent. The respondent must select an answer from available choices. The major advantage of close-ended questions is that responses are easily quantified for tabulation. This greatly expedites analysis. The most common application of close-ended questioning involves scaling techniques. Responses are categorized on a scale between two extremes. Examples of such questions are:

1. Do you receive your inventory reports on schedule?

☐	☐	☐	☐	☐
Never	*Almost Never*	*Usually*	*Almost Always*	*Always*

2. Is the information in your inventory reports accurate?

☐	☐	☐	☐	☐
Never	*Almost Never*	*Usually*	*Almost Always*	*Always*

A final point about close-ended questioning is that only one question should be asked at a time. For example, asking if an inventory report is accurate and on schedule in one question places the respondent in an awkward position if the report is on time, but is not accurate.

Application of Analysis Activities

The techniques of documentation, observation, interviews, and questionnaires are used in varying degrees during the following steps of analyzing the existing system:

1. Review object system.
2. Define decision making associated with object system.
3. Analyze existing information system support of decision making.
4. Isolate deficiencies in the information system.

These steps of systems analysis are discussed in the remaining sections of this chapter.

Review Object System

During this step, project team members have the opportunity to develop a working knowledge of the physical processes associated with the object system under consideration. A clear understanding of the objectives of the object system and of the means used to achieve the objectives should be developed.

The systems analyst should make a special effort to become familiar with the vocabulary associated with the object system. This minimizes communication problems as the systems analysis progresses.

Some type of procedures manual that describes the operation of the object system is usually available. A copy of the manual should be requested from the manager(s) responsible for the object system(s). After the manual has been reviewed, a tour that allows initial observation of the object system in operation should be arranged. Interviews and, if necessary, questionnaires can be used to complete the review.

Define Decision Making Associated with Object System

After an operating knowledge of the object system has been acquired, the next logical step is to define the decision making associated with managing the object system. A definition of the decision-making system (i.e., decision system) provides the framework for determining what information is required. This is one of the most neglected aspects of systems analysis. Since the utility of information is its ability to improve decision making, such negligence is somewhat surprising.

Managers are frequently asked what information they would *like* to have or are offered copies of reports that are currently being produced or will be produced for other managers. This approach tends to encourage managers to ask for more information than they *need.* Research in the area of decision making and the use of information indicates the following points:

1. Decision makers tend to ask for and feel more comfortable with more detailed information than they need. However, they appear to make better decisions with summarized information and exception reporting.[1]
2. The less knowledgeable decision makers are about the decisions required to properly manage a process, the more information they tend to request (presumably, hoping to find something of value). However, much of the information they request is irrelevant to their decision making.[2]

---

1. Norman L. Chervany and Gary W. Dickson, "An Experimental Evaluation of Information Overload in a Production Environment," *Management Sciences*, June 1974, pp. 1335–1344.

2. Izak Benbasat and Roger G. Schroeder, "An Experimental Investigation of Some MIS Design Variables," *MIS Quarterly*, Mar 1977, pp. 37–47.

By basing information requests on the decisions they have to make, managers can be discriminating in their information requests. This reduces the tendency of managers to create information overloads for themselves. Such overloads are both dysfunctional and expensive.

Difficulty in Defining Decision System. Though defining the decision system is a desirable step of systems analysis, it is not a trivial process. Considerable discipline and effort on the part of managers may be required to define such a system. Managers often make decisions so routinely that it seldom occurs to them how often they make decisions and what types of decisions they make. Consequently, they are understandably tempted to take a "shotgun" approach to defining their information requirements rather than a more time-consuming, but more specific, "rifle" approach.

Another difficulty with getting managers to define their information requirements is that the idea of defining and documenting major decisions is not always attractive to managers. Revealing such information may threaten their value to the organization. This is a behavioral consideration; as such, it is discussed in Chapter 9, "Administration of Systems Analysis."

Decision-Making Process. In order to define decision-making processes, it is necessary to understand how decisions are made. The most useful and well-known model of decision making is that proposed by Herbert Simon.[3] His description of a three-phase process is summarized in Table 5-2.

The transaction processing, information providing, decision support, and programmed decision-making functions of an infor-

Table 5-2. *Phases of Decision Making*

Phase of Decision Making	*Explanation*
1. Intelligence	Recognizing a problem or an opportunity that calls for a decision. Gathering and structuring information for enlightenment about the problem or opportunity.
2. Design	Developing and analyzing alternative solutions.
3. Choice	Selecting a particular solution from those available for implementation.

3. Herbert A. Simon, *The New Science of Management Decision,* 1960, p. 54.

mation system identified in Chapter 3 are used to support each phase of the decision-making process. During the intelligence phase, the information system may call problems or opportunities to the decision maker's attention. Once identified, information relevant to the problem or opportunity may be gathered from the information system in both predetermined and ad hoc ways. That is, normally produced reports or special studies generating new reports may be used. These studies may use existing transaction data, or special data (e.g., results of customer surveys) may be required.

The design phase may be supported by the decision-support capabilities of the information system. Decision-support systems allow the decision maker to play "what if" with respect to the decision under consideration. This is accomplished by using a model of the object system. The decision maker can test different decisions using different assumptions. The model then generates the likely outcomes. The design and use of such models are covered in Chapter 11.

Once a decision has been made, the information system provides information based on transaction processing or special studies as feedback to management about the effectiveness of the decision. In some cases, decisions may have to be made on an ongoing and routine basis. If the decision is well structured and defined, it can become a programmed decision, automatically handled by the information system. For instance, inventory can be reordered automatically when stock is reduced to a certain level.

The organizational concepts discussed in Chapter 3 are useful in explaining the difficulty of defining decision-making processes. The more closed/stable/mechanistic the operation of an organizational process, the easier it is to define the decision-making process and also the greater the potential for programmed decision making. The more open/adaptive/organic the organizational process, the more difficult it is to define the decisions that have to be made. Consequently, it is more difficult to predetermine information requirements. An information system in an open/adaptive/organic environment should have great flexibility in providing information and should place great emphasis on decision-support capabilities.

Decision Centers. The decisions made in an organization tend to be clustered into decision centers. A decision center generally consists of a decision maker(s), decision procedures, and an activity for which decisions must be made.[4] Accordingly, the decisions made in

4. Sherman C. Blumenthal, *Management Information Systems: A Framework for Planning and Development,* 1969.

a decision center tend to pertain to the management of a particular organizational process.

Viewing the organization in terms of decision centers is particularly useful in systems analysis. Decision centers are potential areas where organizational processes may be improved by the provision of more relevant, timely, accurate information. In organizational terms, a decision center combined with an activity center constitutes a functional unit or department.

Table 5-3 assumes that an existing order processing system is being analyzed. It shows the decision centers and major decisions likely to exist in the organization.

Another advantage of defining decision centers and major decisions is that doing so provides an overall profile of decision making and where decisions are made. This profile often reveals discrepancies in decision making. For example, if salespersons are under the impression that they can, but they really cannot, this discrepancy should be cleared up. Otherwise, inventory may be committed to customers in error.

If the same decision is being made by two or more decision centers, this too should be resolved. Consider, for instance, a situation where all orders must be approved for credit by the credit department before they are sent to the order processing department. However, the order processing department is unaware of this procedure so a second credit check is made, using an exact copy of the credit rating report used by the credit department. Such duplication of effort is not uncommon in organizations. It may go on, unnoticed, until identified by someone aware of what both departments are doing. Requiring that information received by a department be justified for decision making should expose discrepancies and duplication of decision-making processes.

Analyze Existing Information System

This activity consists of analyzing the existing information system and its support of the decision system. Specifically, this analysis involves determining the collection, processing, storing, and reporting within the existing system.

Some form of documentation of the existing information system is generally available and is the logical starting point for analysis. Perhaps the most useful form of documentation for reviewing an information system is a *systems flowchart.* A flowchart is a graphical representation of the system. Flowcharting symbols are used to represent operations, data flow, files, equipment, and so on. Figure 5-2 illustrates the various symbols used on

Table 5-3. *Decision Centers Involved in Order Processing*

Decision Center	*Activity*	*Examples of Major Decisions*
Salespersons	Selling Merchandise	• Which customers to call on • What to sell customers • What is available to sell
Credit Department	Accounts Receivable Management	• Which customers to allow credit to • How much credit to allow • Which customers need past-due notices • Which customers' credit should be discontinued
Ordering Department	Inventory Management	• What inventory to stock • How much inventory to stock • When to reorder stock • When to unload slow-moving inventory • Which customers to allocate available inventory to
Shipping Department	Packing and Shipping Orders	• What merchandise to send to what customers • What orders can be shipped together to save delivery cost

systems flowcharts. Flowchart templates with cutout forms of these symbols can be purchased and used in constructing such flowcharts.

Figure 5-2. *Systems Flowcharting Symbols*

Processing A major processing function. 	**Input/ Output** Any type of medium or data. 
Punched Card All varieties of punched cards, including stubs. 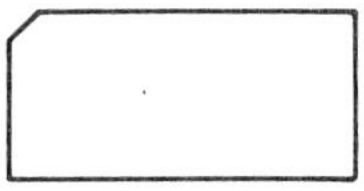	**Perforated Tape** Paper or plastic, chad or chadless. 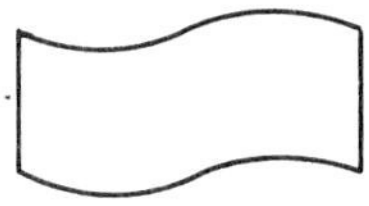
Document Paper documents and reports of all varieties. 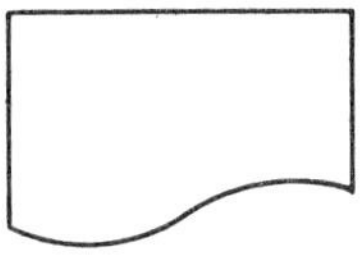	**Transmittal Tape** A proof or adding-machine tape or similar batch-control information. 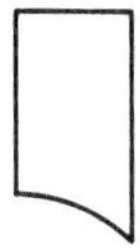
Magnetic Tape 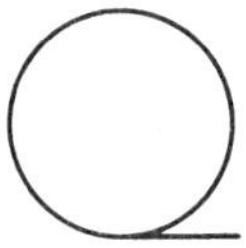	**Disk, Drum, Random Access** 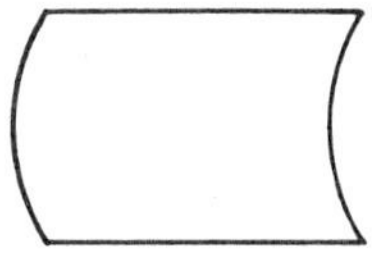
Offline Storage Offline storage of either paper, cards, magnetic tape, or perforated tape. 	**Display** Information displayed by plotters of visual-display devices. 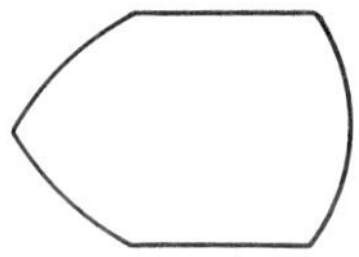
Online Keyboard Information supplied to or by a computer utilizing an online device. 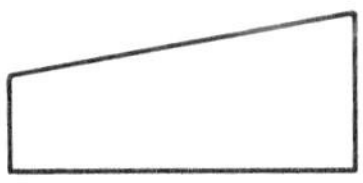	**Sorting, Collating** An operation on sorting or collating equipment. 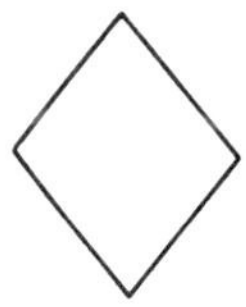
Keying Operation An operation utilizing a key-driven device. 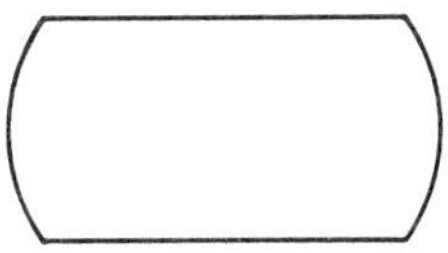	**Clerical Operation** A manual offline operation not requiring mechanical aid. 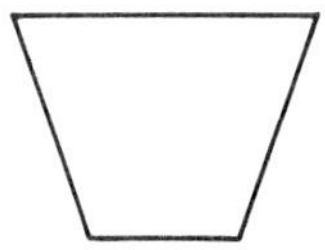
Auxiliary Operation A machine operation supplementing the main processing function. 	**Communication Link** The automatic transmission of information from one location to another via communication lines. 
Flow The direction of processing or data flow. 	

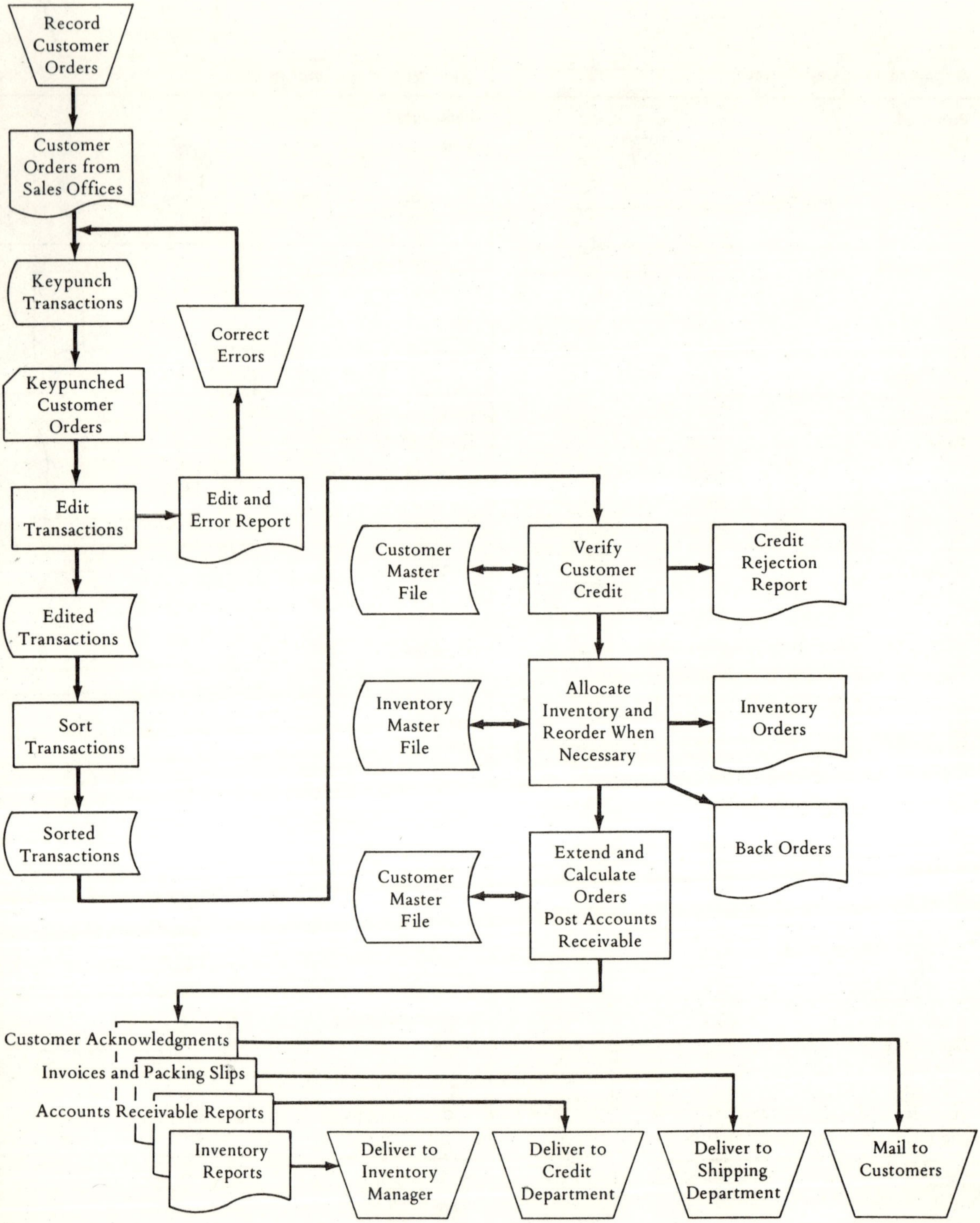
Record Customer Orders
Customer Orders from Sales Offices
Keypunch Transactions
Correct Errors
Keypunched Customer Orders
Edit Transactions
Edit and Error Report
Edited Transactions
Sort Transactions
Sorted Transactions
Customer Master File
Verify Customer Credit
Credit Rejection Report
Inventory Master File
Allocate Inventory and Reorder When Necessary
Inventory Orders
Back Orders
Customer Master File
Extend and Calculate Orders Post Accounts Receivable
Customer Acknowledgments
Invoices and Packing Slips
Accounts Receivable Reports
Inventory Reports
Deliver to Inventory Manager
Deliver to Credit Department
Deliver to Shipping Department
Mail to Customers

Flowcharts can be used to convey varying degrees of detail. In Chapter 1, a basic overview flowchart of an order processing system was presented (see Figure 1-2). A more detailed flowchart of the same system is provided in Figure 5-3. It is important to note that Figure 5-3 represents only the order processing system (subsystem) of an information system. The order processing system uses the inventory master file in the processing of orders. However, the order processing flowchart does not include all the processing necessary to control inventory. Another subsystem is required to maintain (add, delete, or modify) the inventory master file. This subsystem is illustrated in Figure 5-4. The inventory master file updated in the inventory subsystem is the file that is used in order processing.

Flowcharts may not exist for poorly documented systems, or for systems that have not been implemented on a computer system. In such cases, flowcharts must be developed from scratch, from information gathered by observing the operation of the system and by interviewing various persons involved in the operation of the system.

The systems flowchart indicates the various reports, files, and transaction documents used in an existing system. (If the system is manual, the files may be in filing cabinets.) Collecting and analyzing copies of the various reports, file contents, and transaction documents in the context of the systems flowchart leads to a thorough understanding of the actual processing involved in transforming data into usable information.

Isolate Deficiences in the Information System

The final step of analysis involves contrasting the decision system with the information generated from the information system. The objective of this contrasting is to identify discrepancies between the information required and the information available. Deficiencies in the information system can then be identified.

In Chapter 1, deficiencies in information systems were categorized as inclusion and/or structure deficiencies. Inclusion deficiencies pertain to *what* information, technology, and personnel are included or lacking in the system. Structure deficiencies pertain to *how* the information, technology, and personnel are organized and interrelated throughout the system. (To review this framework, refer to Table 1-1.)

Figure 5-3. *Detailed Flowchart of an Order Processing System*

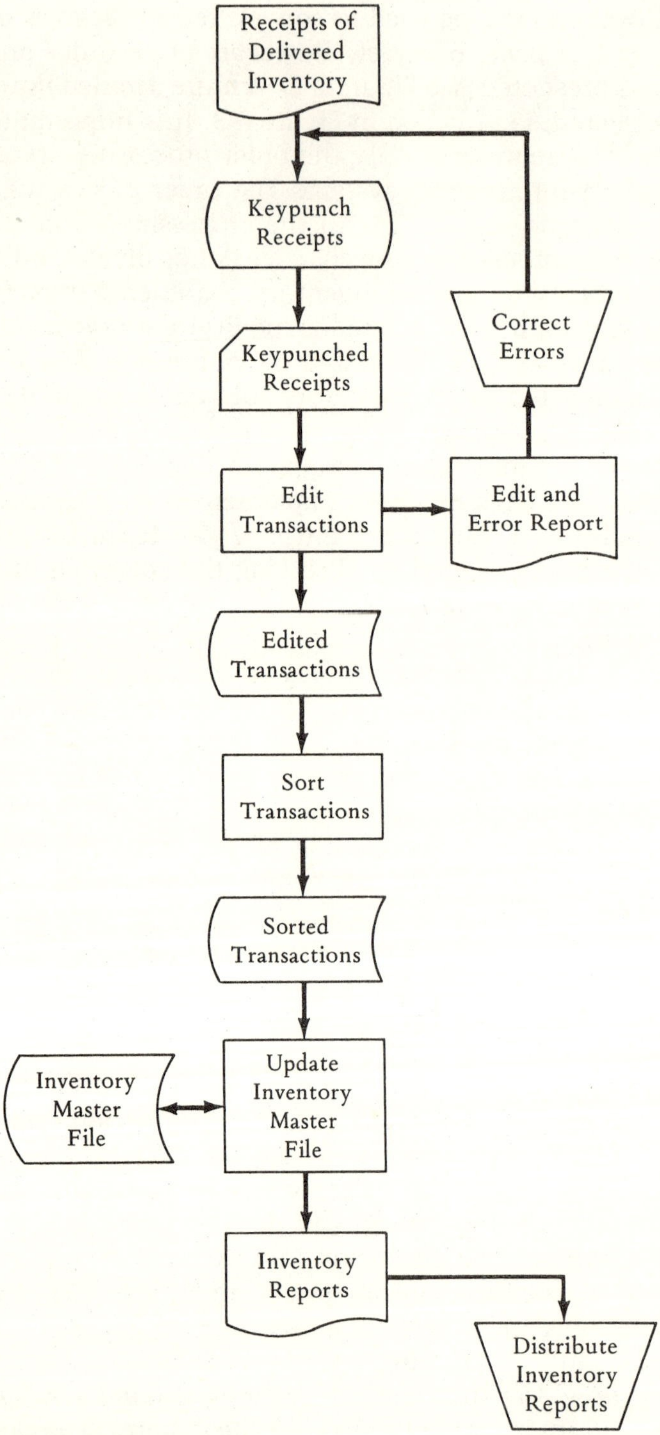

Figure 5-4. *Flowchart of an Inventory Subsystem*

Information Deficiencies

The decision makers, of course, must decide what information is relevant to the decisions associated with their management responsibilities. For example, the manager responsible for inventory may say that the following information is necessary for reordering decisions:

1. Date-of-last-order
2. Quantity-on-hand
3. Quantity-on-order
4. Back-orders
5. Date-of-last-purchase order
6. Sales of item
   a. This month
   b. One month ago
   c. Two months ago
   d. Three months ago
   e. Year-to-date
7. Suppliers
8. Suggested order size

Given the preceding information requirements, consider the following inventory information system deficiencies:

1. Information on back orders and date of last purchase order are not currently processed.
2. Information on sales of item last year is provided, but it is not used in decision making.
3. Necessary information is provided, but it is spread through several reports and is, therefore, time-consuming to retrieve.
4. Year-to-date sales information is processed, but it is not distributed to the manager responsible for inventory.
5. The timeliness of inventory information is inadequate. Because the existing computer system is too small to handle the daily workload, inventory reports have to be processed over the weekend.
6. Inventory information is stored on magnetic tape. Consequently, inventory transactions that arrive after the inventory file is updated but prior to printing of reports are not processed before the printing. (The entire inventory file would have to be processed to update a few records.) Therefore, inventory reporting on those records is not current until the next processing and reporting cycle.
7. Inventory transactions are keypunched but, due to lack of control, verification is often bypassed. This results in frequent errors in inventory information.
8. The inventory program contains a "bug" that results in erroneous computation of quantity on hand.

9. The decision maker is unqualified to use information produced by the system.
10. The programmer who works on the system is incompetent.
11. The least-experienced keypunch operators are verifying inventory transactions, and the error rate is high.
12. The manager of the order processing department is not defining, encouraging, or rewarding good performance by order processing clerks. Both the productivity and the morale of the department are low. This leads to a high error rate and delayed processing.

These information system deficiencies can be categorized individually as inclusion deficiencies and structure deficiencies as illustrated in Table 5-4.

PIECES

In the discussions of problem and opportunity identification and definition at the first of this chapter, it is indicated that management's expression of problems or opportunities can generally be categorized into the PIECES framework. For thorough analysis, PIECES problems or opportunities must be translated into inclusion and/or structure deficiencies. For example, the performance of an object system may be poor for one or more of the following reasons:

1. Information that could improve performance
   a. is not processed by the information system. (Information Inclusion)
   b. is in the wrong form. (Information Structure)
   c. cannot be processed fast enough. (Technology Inclusion)
   d. is computed incorrectly. (Technology Structure)
2. Decision maker is unqualified. (Personnel Inclusion)
3. Personnel are unaware of how to properly process information. (Personnel Structure)

Isolation of the specific cause(s) of a system deficiency requires acute perceptive skills on the part of those involved in systems analysis. They must avoid improperly analyzing the cause of the problem or preventing an opportunity. A problem or an opportunity must be continually refined until it can be uniquely identified and categorized. The completed analysis then sets the stage for systems design, which is covered in the next chapter.

Table 5-4. *Categorization of Inventory Information System Deficiencies*

		Inclusion Deficiency	*Structure Deficiency*
INFORMATION	Definition	Appropriate information is lacking, and/or inappropriate information is included in the system.	The manner in which information is collected, stored, and/or reported in the system is difficult or inconvenient to utilize.
	Information System Deficiencies*	1 and 2	3 and 4
TECHNOLOGY	Definition	Appropriate technology is lacking, and/or inappropriate technology is included in the system.	The hardware and software included in the system are not organized to most effectively and efficiently process the available information.
	Information System Deficiencies*	5 and 6	7 and 8
PERSONNEL	Definition	Appropriate personnel are lacking, and/or inappropriate personnel are included in the system.	The personnel associated with the system are poorly organized and/or managed.
	Information System Deficiencies*	9 and 10	11 and 12

* Numbers 1 through 12 refer to the inventory information system deficiencies listed on pages 109 and 110.

SUMMARY

Problems and opportunities are identified by either reaction or opportunistic surveillance. They may exist primarily in either the object system or the information system. Though the systems analyst is primarily concerned with the information system, systems analysis may also involve the object system.

Problems and opportunities can be broadly categorized as related to performance, information, economy, control, efficiency, and security (PIECES). A particular problem or opportunity may belong within one or more of the PIECES categories.

Minor problem analysis and resolution may require little more than a quick review of the situation and a memo to an appropriate manager. More complex problems are usually addressed by first conducting a feasibility study and then organizing a project team to do detailed analysis.

The fundamental activities for analyzing a system consist of reviewing documentation, observation, interviews, and questionnaires. The four steps of the analysis are:

1. Reviewing object system
2. Defining decision making associated with object system
3. Analyzing existing information system support of decision making
4. Isolating deficiencies in the information system

The result of the analysis defines inclusion and structure deficiencies as they pertain to the information, technology, and personnel aspects of the information system.

EXERCISES

1. Differentiate between management actions that are reactionary and those that are opportunistic in orientation. Give examples of each.
2. Discuss the difference between an object system and an information system.
3. Refer to Table 3-7. Discuss how an information system can be used to improve the operation of the various mainstream processes. (A mainstream process is an object system.)
4. Identify and discuss how the different categories within PIECES (performance, information, economy, control, efficiency, and security) apply in the following situations. Remember that more than one category can apply. If additional information is required to more accurately pinpoint the problem, discuss this also.
   a. Employees are gaining unauthorized access to payroll information.
   b. Due to insufficient computer capacity, the response time for online credit checks is too slow.
   c. The cost of running the delivery department has increased considerably with no increase in service.

- **d.** Salespersons do not receive timely enough information to determine if they have adequate inventory to satisfy customer orders.
- **e.** Poorly constructed products are getting past quality control and being shipped to customers.
- **f.** Important information on quality control activities (i.e., number of products checked, rejection rates, etc.) is not collected or processed.
- **g.** Though information on customer opinions of product quality is collected and processed, it is not made available to the production manager.

5. Discuss the fundamental activities of analysis (reviewing documentation, observation, interviews, and questionnaires) as they may be used for:

- **a.** Reviewing the object system.
- **b.** Defining decision making associated with an object system.
- **c.** Analyzing existing information system support of decision making.
- **d.** Isolating deficiencies in the information system.

6. Develop a systems flowchart for a payroll system described as follows:

- **a.** Payroll transactions (employee data, pay rate, hours worked, etc.) are prepared by departments and sent to the payroll department.
- **b.** The payroll department reviews payroll transactions and sends them to the data processing department for keypunching.
- **c.** Transactions are keypunched.
- **d.** Keypunched transactions are validated (checked for errors) by a computer program that creates an error report and also a validated transaction file on magnetic disk.
- **e.** The error report is sent back to the payroll department for corrections. Corrected transactions are then sent back to the data processing department, and steps *c* and *d* are repeated.
- **f.** When all errors have been corrected, the validated transaction file is sorted into social security number order by a computer sort program. It is then inputted along with the payroll master file (also a disk file) to a computer program that updates the payroll master with new employee data and then performs the payroll computations (e.g., multiplies hours worked by pay rate). The same program prints the following payroll output:

  Paychecks
  Pay register
  Check reconciliation journal

- **g.** The paychecks are sent to employees. The other documents are sent to the payroll department.

7. Situations a, b, d, f, and g in Exercise 4 clearly pertain to deficiencies of the information system. (Situations c and e appear to be problems with the object system.) Using a system deficiency table (see Table 5-4), categorize the information system deficiencies.

8. Using a system deficiency table (see Table 5-4), categorize the problems and opportunities on page 111. Remember that personnel deficiencies caused by negligence, lack of training, poor morale, and the like are structure deficiencies. Personnel deficiencies caused by personnel not being capable of doing a job are inclusion deficiencies.

SELECTED REFERENCES

Ackoff, Russell L. "Management Misinformation Systems," *Management Science,* Dec 1967, pp. 147–156.

Anderson, William S. "Bridging the Systems Expectation Gap," *Data Management,* Nov 1975, pp. 29–32.

Argyris, Chris. "Resistance to Rational Management Systems," *Innovation* No. 10, 1970, pp. 28–35.

Benbasat, Izak, and Roger G. Schroeder. "An Experimental Investigation of Some MIS Design Variables," *MIS Quarterly,* Mar 1977, pp. 37–47.

Blumenthal, Sherman C. *Management Information Systems: A Framework for Planning and Development.* Englewood Cliffs, N.J.: Prentice-Hall, 1969.

Chervany, Norman L., and Gary W. Dickson. "An Experimental Evaluation of Information Overload in a Production Environment," *Management Sciences*, June 1974, pp. 1335–1344.

Davis, Gordon B. *Management Information Systems: Conceptual Foundations, Structure, and Development.* New York: McGraw-Hill, 1974.

Dickson, Gary W., and John K. Simmons. "The Behavioral Side of MIS," *Business Horizons,* Aug 1970, pp. 59–71.

Feltham, Gerald A. "The Value of Information," *The Accounting Review*, Oct 1968, pp. 684–694.

Hirsch, Rudolph E. "The Value of Information," *The Journal of Accountancy*, June 1968, pp. 41–45.

Johnson, Richard A.; Fremont E. Kast; and James E. Rosenzweig. *The Theory and Management of Systems.* New York: McGraw-Hill, 1973.

Lucas, Henry C., Jr. "A Descriptive Model of Information Systems in the Context of the Organization," *Data Base,* Winter 1973, pp. 27–39.

Nolan, Richard L. "Systems Analysis for Computer-Based Information Systems Design," *Data Base,* Winter 1971, pp. 1–10.

Nugent, Christopher E., and Thomas E. Vollman. "A Framework for the Systems Design Process," *Decision Sciences,* 1972, pp. 84–109.

Simon, Herbert A. *The New Science of Management Decision.* New York: Harper & Brothers, 1960, pp. 54ff.

Wetherbe, James C., and Carlton J. Whitehead. "A Contingency View of Managing the Data Processing Organization," *MIS Quarterly,* Mar 1977, pp. 19–25.

6 Systems Design For Information

This chapter presents the concepts and techniques that are the core, and perhaps the glamorous aspect, of systems analysis—the design of information systems. During the design stage, the systems analyst begins designing the means of solving the problems or capitalizing on the opportunities discovered during the analysis stage. The systems analysis stage defines the way things are; the systems design stage defines the way things should be.

FRAMEWORK FOR SYSTEMS DESIGN

Systems design is a process wherein the inclusion and structure aspects of the information system are redesigned in an effort to align more closely the information system with the information requirements of the organization. Though the systems analyst plays a key role in performing this task, he or she does not singlehandedly accomplish it. Table 6-1 illustrates the roles and relationships of those involved in systems design with respect to the inclusion and structuring of revised or new information systems.

Information

Managers must first decide what information needs to be provided by an information system. Systems analysts assist in this activity by reviewing decisions made and information that can be made available. Once the information to be provided by the system has been determined, systems analysts play the lead role in structuring the information in terms of its collection, storage, and reporting.

Technology

The selection of the technology to be applied in processing is generally a joint effort between managers and systems analysts. Computer vendors are called upon to provide information on current hardware and software available. The structuring of the technology is primarily a computer programming activity in which systems analysts assist by providing detailed specifications as required.

Table 6-1. *Roles of Systems Design Participants*

	Inclusion	*Structure*
INFORMATION	1. Manager(s) 2. Systems Analyst(s)	1. Systems Analyst(s) 2. Manager(s)
TECHNOLOGY	1. Manager(s) and Systems Analyst(s) 2. Computer Vendor(s)	1. Programmer(s) 2. Systems Analyst(s)
PERSONNEL	1. Manager(s) 2. Systems Analyst(s)	1. Manager(s) 2. Systems Analyst(s)

Legend: 1 = Lead role.
2 = Support role.

Personnel

Managers generally make the decisions pertinent to the selection and structuring or organization of personnel. However, systems analysts are often called upon to make recommendations as to personnel and their training.

Structuring Information

The structuring of information in the information system is the primary domain of the systems analyst and the major thrust of this chapter. The inclusion and structuring of technology and personnel that represent the means to support the information structure are discussed in Chapters 7 and 8.

SYSTEMS DESIGN AND SPECIFICATION

When structuring information for an information system, a systems analyst may not know what specific hardware and software will be best suited to support the system. Therefore, systems design techniques and procedures should be as hardware and software independent as possible. This allows the systems specifications to be implemented using whatever hardware and software turns out to be most cost effective in supporting the information structure. The relationship between information structure design and hardware and software is similar to the relationship between architectural design and building materials. In both cases, the specifications should articulate with sufficient precision what the final product should be without unduly restricting the selection of the best available means for constructing the final product.

Achieving a balance between design techniques that are both hardware/software independent and yet specific enough to be practical presents somewhat of a dilemma to systems analysts. In this book, an improvised and updated version of a technique entitled *Accurately Defined Systems* (ADS) is used as a tool for designing systems specifications.[1] This technique is a means of developing precise systems specifications without unduly imposing restrictions concerning particular hardware and software.

The process of structuring information involves the following systems design specifications:

1. **Output Definitions.** Describe the printouts or terminal displays to be provided by the system.
2. **Input Definitions.** Display the format of each data field or element coming into the system.

1. Special appreciation is extended to the National Cash Register Company for granting permission for the use of material taken from *Accurately Defined Systems,* © copyright 1968, by the National Cash Register Company, Dayton, Ohio, U.S.A.

3. **Data Element Dictionary.** Defines all data fields that are to be inputted, computed, stored, and reported.
4. **Decision Tables.** Illustrate complex logical relationships necessary for input, computation, and reporting of information.
5. **Systems Flowchart.** Provides a graphical representation of the system, using symbols to represent operations, data flow, files, equipment, and so on.

The various design specifications are linked together such that the input, computation, logical processing, storage, and reporting of each item of information can be traced forward or backward from its point of origin to its final use(s). Upon completion, the specifications should provide sufficient documentation to guide the computer programming required.

Though organizations may have systems design procedures that differ from the procedures presented in this chapter, the differences will generally be cosmetic rather than conceptual in nature. Readers who master the techniques presented in this chapter should experience a comfortable transition to other systems design procedures.

Encoding and Compression

Before the five types of design specifications noted above are described in detail, a brief discussion is necessary to explain how techniques of data encoding and data compression are used to save time and space in defining data elements, report headings, and so forth.

The form in which information is used external to computers is generally more lengthy than is necessary or desirable for machine processing. Reducing the number of characters or digits required to represent information results in reduced storage and processing costs. It also saves space on transaction documents and reports.

Encoding and compression are two basic approaches to represent information efficiently. *Encoding* consists of translating data from one form to another. For example, job titles can be coded as numbers. *Compression* consists of eliminating characters that are not essential to the information meaning. Abbreviation is the most common compression technique; it is effectively achieved by the removal of vowels from information or the truncation of trailing characters. Examples of encoding and compression are given in Tables 6-2 and 6-3, respectively.

Table 6-2. *Encoding of Job Titles*

Job Title	*Code*
Programmer	100
Computer Operator	110
Systems Analyst	120
.	.
.	.
.	.
Accountant	250
Auditor	260
Controller	265

Frequently, information must be added to a table of coded information. Therefore, it is often desirable to leave gaps between codes to facilitate convenient insertion of new information. For example, in Table 6-2, a new position of Senior Programmer can bc coded as 105.

Coding generally requires less space than compression. However, codes often have to be translated back to the original representations to be useful external to the computer. Compressed data can generally be understood without translation.

Output Definitions

The output from an information system consists primarily of printed reports and/or displays on a CRT (cathode ray tube) terminal. The types of information on the various reports and displays

Table 6-3. *Compression of Merchandise Description*

Merchandise	*Compressed*
Flashlight	Flshlght
Battery	Bttry
Hammer	Hmmr
Charcoal Grill	Chrcl Grll
.	.
.	.
.	.
Briquettes	Brqtts

are determined by the analysis of information requirements as discussed in Chapter 5.

The definition of output involves laying out the formats for the various reports and/or displays. Report layout forms are used to define the specific location of each character to be printed (see Figure 6-1). Layout forms generally contain 132–160 horizontal print positions and 30–60 vertical print positions. However, when possible, the target hardware should be considered when designing a report or display. For example, if a report is to be printed on a 132-character printer, then not more than 132 horizontal print positions should be used for one line. If output is to be displayed on a CRT, only the number of character positions available on that device should be used.

The actual format that should be used for a report or display is to a great extent a matter of taste. As a frame of reference, the systems analyst can review any existing reports or displays currently produced and/or ask persons who will use the reports or displays for their preferences as to output format.

Headers and Information

Reports or displays contain both headers and information. Headers are the report titles, column headings, page numbers, and dates associated with a report or display. Information is the actual content of the output.

Headers may be printed by the computer or preprinted on custom-made forms. In either case, the headers are formatted on the layout form by positioning the letters of the headers using column and line numbers.

After the headers are formatted, the format of the information to be printed under the headers can be layed out. The symbols in Table 6-4 are used to convey, in a standard fashion, the forms in which information fields are to appear on the report. Information fields that are printed on successive lines need only have print symbols shown for the first line. A wavy vertical line drawn under a field is sufficient to convey that the field is to be repeated on successive lines (see Figure 6-2).

The conditions causing each line of a report or a display to be printed should also be conveyed on the report layout. The condition causing a line to be printed is generally a change in the values of certain information fields, or the start of a new page or display. For example, Figure 6-3 illustrates a report on which totals are printed. The report is to have a detail header at the top of every page with successive lines of detail information below it. When the

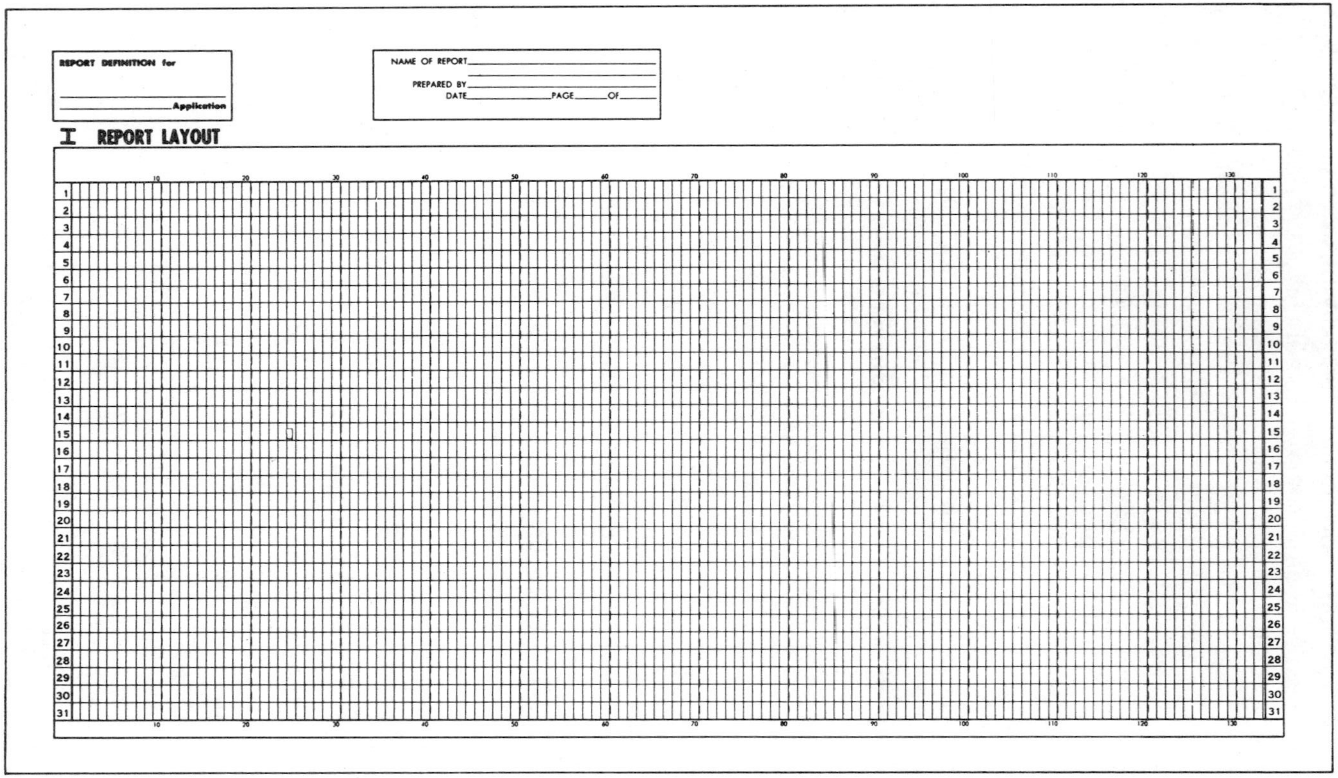

Figure 6-1. *Layout Form for Output Definition*

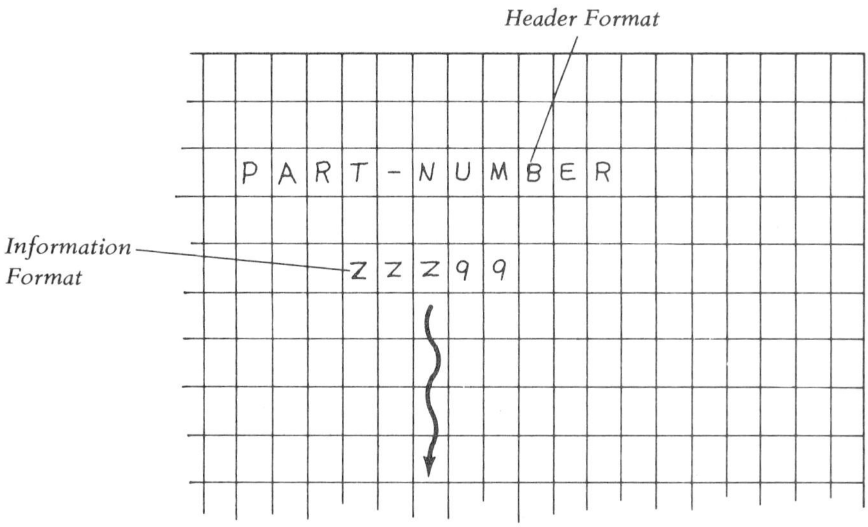

Figure 6-2. *Formatting Headers and Information*

content of the department field changes, a department total is printed. The department field is then reset to zero and incremented during the totaling of the detail for the next department. A store total is printed when there is a change in the content of the store field. Since the department field changes whenever the store field changes, it is also printed. Both department and store are then reset to zero, and processing resumes. A corporate total is printed at the end of the report. It represents the grand total of all departments and stores.

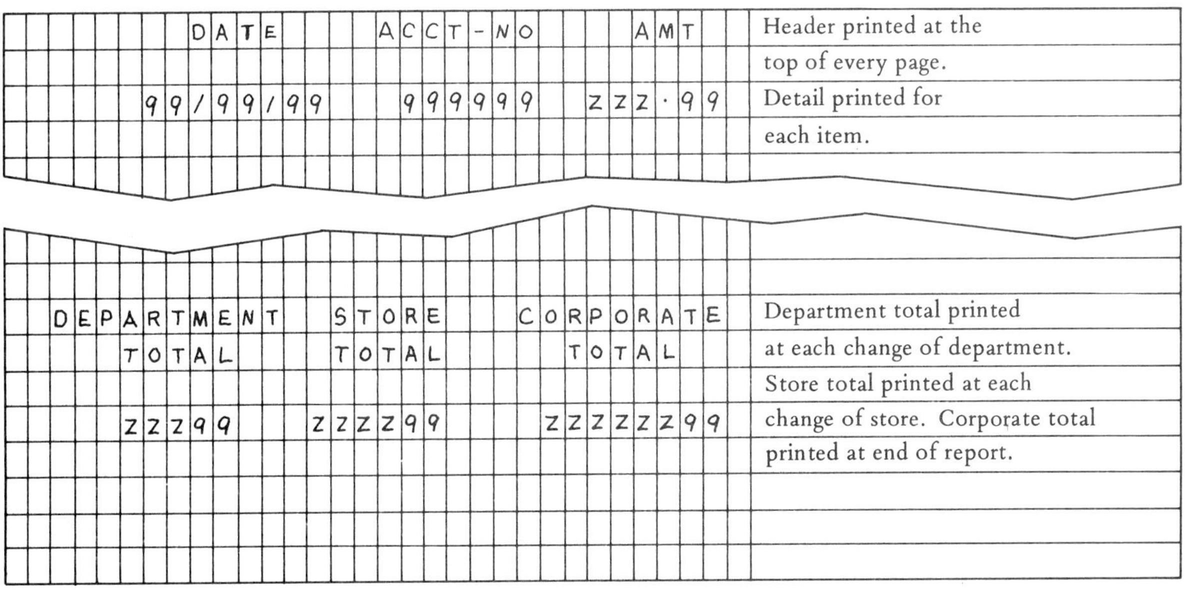

Figure 6-3. *Documenting Conditions under Which Report Lines Are to Be Printed*

Table 6-4. *Symbols Used to Format Information Fields*

9 = Numeric
A = Alphabetic
X = Alphanumeric (characters or numbers)
Z = Zero suppress a numeric field
* = Lead numeric field for dollar protection
B = Embedded blank in an alphanumeric field
$ = Floating dollar sign if substituted for Z
. = Decimal point

Examples of formats followed by sample data.

									9	9	9	9
									0	0	3	2
									4	0	6	9

- Four-digit numeric field without zero suppression. Any numeric digit (0–9) may appear in any of the four positions.

									Z	9	9	9
										9	8	7
									4	6	2	5

- Four-digit numeric field with zero suppression for the first character only. If the high-order position is zero, it appears as blank; all other digits are printed. The last three positions are printed regardless of value.

							X	X	X	X	X	X
								J	O	N	E	S
							W	I	L	S	O	N

- A six-character alphanumeric field. The numbers 0-9, all letters of the alphabet, and special symbols (i.e., %, *, ?, etc.) can appear in this field.

Linking Output Definitions

To interface output definitions to other design specifications requires some type of linking. The systems analyst must indicate the source of all information that is to appear on a report or display. To accomplish this, each field of information must be assigned a unique (and preferably meaningful) name. That unique name is then used to refer to that field of information or data throughout all design specifications. It is imperative that the systems analyst be consistent in both the unique identity and the spelling of each name. Otherwise, confusion and erroneous processing are likely.

X	X	B	9	9
A	B		3	2
B	6		0	4
3	0		5	0

- Five-position field containing two alphanumeric characters and two numeric characters separated by a blank. The numerals are not zero suppressed.

$	$	$	$	9	.	9	9
		$	4	9	.	5	0
$	6	8	9	0	.	0	0
			$	0	.	5	0

- Eight-position numeric field with zero suppression on the first four positions. Multiple dollar signs indicate a floating dollar sign is to replace the right-most insignificant zero. All other insignificant zeroes are suppressed. The high-order digit should always be zero to insure the printing of a dollar sign.

$	*	*	*	.	9	9
$	*	*	8	.	5	0
$	1	6	8	.	3	5
$	*	8	5	.	0	0

- Seven-position field with a fixed dollar sign. The asterisk (*) indicates the use of dollar protection by placing leading asterisks in unused dollar positions.

The names of the data fields shown on an output definition should be numbered and listed on a separate sheet of paper attached to the definition. The numbers of the names can then be placed within parentheses, directly under the corresponding data fields on the report or display layout. This provides a cross-reference between the fields printed on the report and the names of the fields (see Figure 6-4).

Though names act as a link to data fields on other design specifications, at this point the other specifications have not been prepared. Therefore, at the time that the report and display definitions are prepared, the source of a data field may not be known.

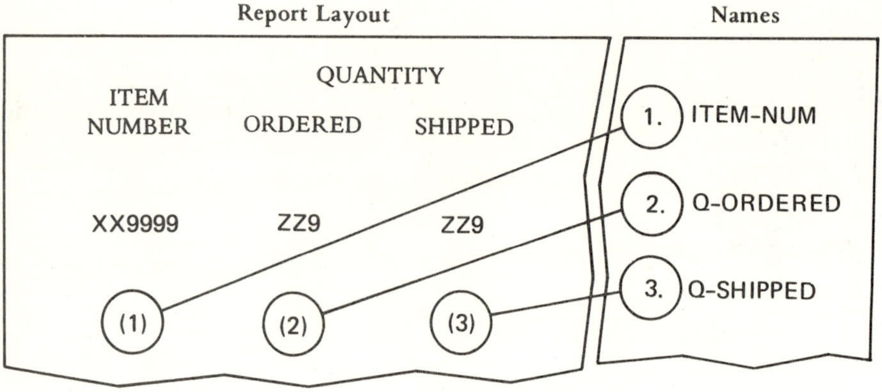

Figure 6-4. *Creating a Cross Reference to Data Fields*

This is appropriate; the output definitions are set up to determine what information is needed. Once the information needed is determined, the other specifications are created to determine what data and what processing are required to generate that information.

Sequencing and Selection

An output definition must also convey any sequencing or selection considerations of the output. *Sequencing* defines what, if any, sorting is needed prior to reporting. *Selection* defines what, if any, items are to be excluded from reporting. Both the sequencing and selection logic of a report can be documented at the bottom of the report layout or on a separate sheet attached to the report.

The sequencing of a report is expressed in terms of sequence levels. For example, if a report is to be printed in department number order, it has one level of sequencing. If a report is to be sequenced by department number within store number, it has a two-level sequence. A sort on department number is required first; it must be followed by a sort on store number. Generally, the highest, or last, level of the sort is called the *major sort.* Any additional lower levels are called *minor sorts.* A sorting sequence can then be expressed as follows:

Output Sequencing:

Major	Level 1.	*Region Number*
	Level 2.	*Store Number*
	Level 3.	*Department Number*

In the preceding example, the information will be sorted on the last specified sort level (department number) first, then on store number, and finally on region number.

Selection logic is only required when not all items of a given set of information are to be reported. In simple cases, the selection can be narratively described as illustrated below:

Output Selection Logic:
Exclude all accounts with amounts less than $1.00

For more complex selection logic, a decision table (discussed later in this chapter) should be used. When a decision table is used to express output selection logic, the output definition need only provide the number and/or the name of the decision table.

Form Identification

One additional item of information necessary to complete an output definition is some type of form identification for documentation and linking purposes. Most layout forms provide space for some type of form identification (see Figure 6-1). The minimum points to be documented on each output definition are illustrated by the following example:

Name of Application or System:	Payroll System
Name of Report:	Payroll Checks
Prepared by:	John Smith
Date:	5/6/80

Input Definitions

An input definition is used to lay out the format of data coming into the system. Standard layout forms can be purchased or designed for this function (see Figure 6-5.) The forms are not hardware oriented and can be used for any type of input medium.

Each input data field is assigned a unique name. This name is then used to refer to the data field throughout all design documentation.

REPORT DEFINITION for

______________________ Application

NAME OF REPORT ______________________

PREPARED BY ______________________

DATE ____________ PAGE ______ OF ______

I REPORT LAYOUT

1 2 3 4 5 6 7 8 9 10 11 12 13 14 15 16 17 18 19 20 21 22 23 24 25 26 27 28 29 30 31 32 33 34 35 36 37 38 39 40 41 42 43 44 45 46 47 48 49 50

NAME OF FORMAT: ______________________

1 2 3 4 5 6 7 8 9 10 11 12 13 14 15 16 17 18 19 20 21 22 23 24 25 26 27 28 29 30 31 32 33 34 35 36 37 38 39 40 41 42 43 44 45 46 47 48 49 50

NAME OF FORMAT: ______________________

1 2 3 4 5 6 7 8 9 10 11 12 13 14 15 16 17 18 19 20 21 22 23 24 25 26 27 28 29 30 31 32 33 34 35 36 37 38 39 40 41 42 43 44 45 46 47 48 49 50

NAME OF FORMAT: ______________________

1 2 3 4 5 6 7 8 9 10 11 12 13 14 15 16 17 18 19 20 21 22 23 24 25 26 27 28 29 30 31 32 33 34 35 36 37 38 39 40 41 42 43 44 45 46 47 48 49 50

The layout form defines the number of character positions in each field, using the numbers at the bottom of each line for field positioning. As many as eighty character positions are available per line on the form. Additional positions can be specified by simply continuing to the next line.

Record Length

Input transactions may be either fixed length or variable length. Illustrations of both formats are provided in Figures 6-6*a* and 6-6*b*.

Under fixed length, each transaction has the same format and length as every other transaction. Under variable length, this need not be the case.

Variable-length transactions are most often used when a field or group of fields may be repeated within a record. For example, when students enroll for classes, the number of courses they register for varies. Accordingly, for each transaction there are fixed fields identifying a student in a standard format followed by variable fields that are repeated for each course for which the student is registered (see Figure 6-6*b*). The number of courses a student is enrolled for is indicated by the NUMBER-COURSES field that precedes the repeatable COURSE-NUMBER field. Notice that it is only necessary to show the layout of a repeatable field one time. The repetition is conveyed by inserting the word *repeats* or a similar comment next to the field.

Multiple Inputs

Many applications accept more than one type of input transaction. For example, a payroll system might accept the following transactions:

- New employee transaction—Indicates employee social security number, name, department, pay rate, date-of-hire, etc.
- Weekly pay transaction—Indicates employee social security number, hours worked, date, etc.

The social security number is used in both transactions as the key, or record identification, to access the employee record stored in the payroll file. However, whether a transaction is a "new employee transaction" or a "weekly pay transaction" needs to be identifiable to the system so that the transaction can be processed correctly. This is accomplished by the use of a transaction code that uniquely defines each type of input transaction. The transaction code can be located in any position of the transaction, but once a position has been determined, it must be used consistently for all transactions inputted to a system. This is necessary since the computer program will first check this code to determine what

Figure 6-5. *Layout Form for Input Definition*

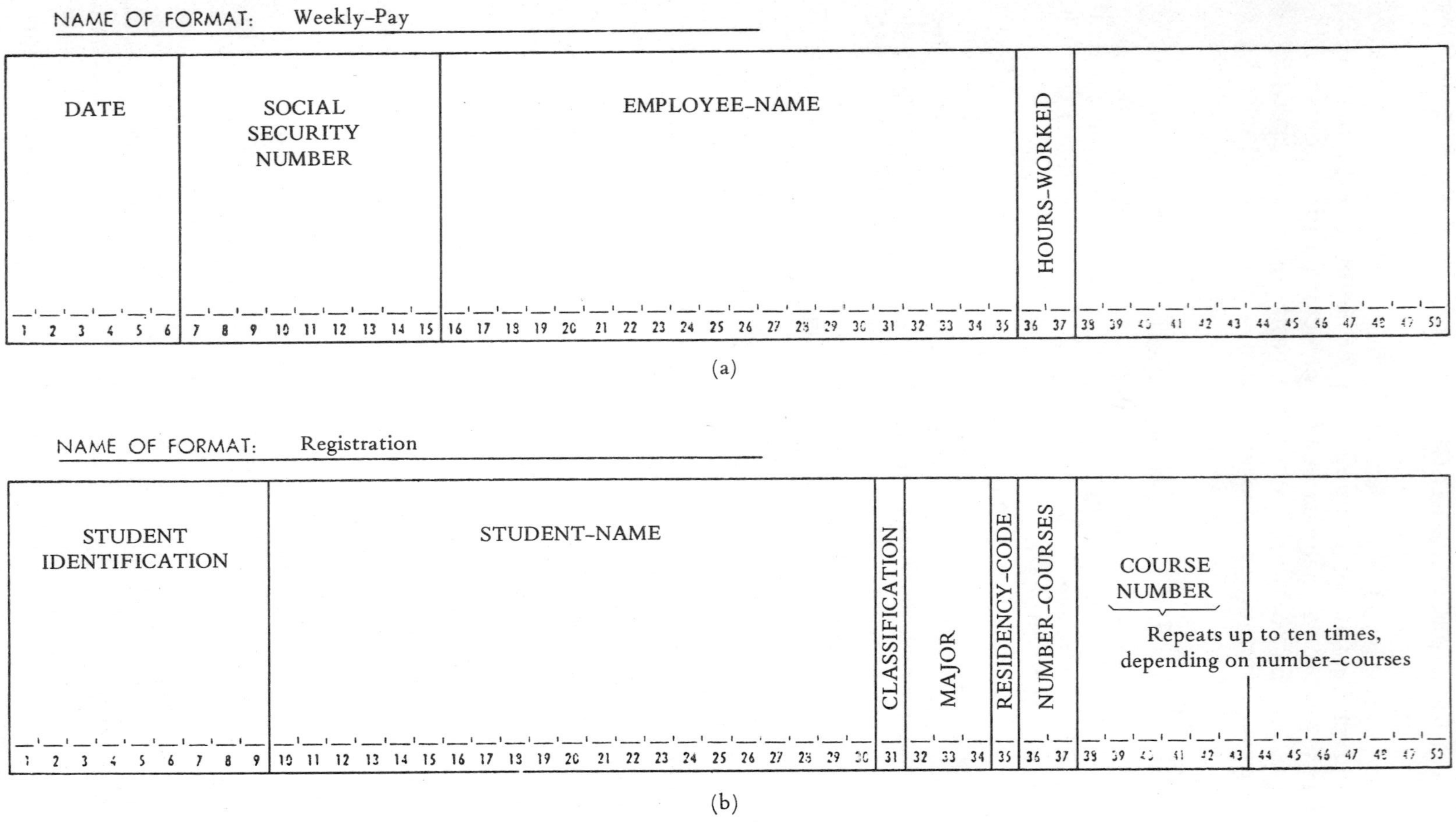

Figure 6-6. *(a) Fixed-Length Transaction. (b) Variable-Length Transaction*

is to be done with a particular transaction. In the case of the two types of payroll transactions, a transaction code could be located in the last two columns of the transaction. The code values, 01 and 02, could identify new employee and weekly pay transactions, respectively.

Linking Input Definitions

An input definition needs to be linked to both the source document from which input data is collected and the remainder of the systems design specifications.

The input definition defines the format in which data is to come into the computer system. However, it is highly unlikely that this format will be used for original data capture. Rather, a more familiar and traditional transaction document format is generally used (say, a customer order form or a bank deposit slip). The contents of these documents are then converted into machine-readable form in the format of the input definition. Accurate cross-referencing of the data fields from the original (source) document to the input definition is, therefore, required. This can be accomplished using a coding scheme similar to the one described for an output definition in which corresponding data fields are numbered identically.

The input definition is linked to the remainder of the systems specifications by the use of a unique name for each data field. As noted earlier, each data field should be referred to in exactly the same way throughout all design documents.

Form Identification

Each input definition should be identified for documentation purposes. An area for this identification is usually provided on the input definition form (see Figure 6-5). The minimum points to be documented on each input definition are illustrated by the following example:

Name of Application or System:	Payroll System
Name of Transaction:	Weekly Pay Transaction
Sequence of Transactions:	Social Security Number
Media Type:	Punched Card
Prepared by:	John Smith
Date:	5/6/80

Data Element Dictionary

The data element dictionary defines every data field in the system. The data element dictionary provides the following information for each data element:

- Name
- Definition
- Source
- Where Used
- Maintenance
- Storage

Name

The data element name is the unique name assigned to a data element. This name can be used to track an element referred to in an output, input, or decision table back to the data element dictionary for a complete description.

Because the data elements will be referred to in computer programs, it is advisable to choose data element names that can be used as names in computer programs. This allows data names in computer programs to be directly cross-referenced to the data element dictionary. Programming languages such as COBOL and PL/I allow the use of lengthy (generally a minimum of twenty-five characters) and meaningful data names. When appropriate characters within names can be separated by hyphens or underscores for clarity. RPG, another commonly used programming language, generally allows only six characters per name.

Through the compression techniques described earlier in this chapter, meaningful data names can be created and used in computer programs. For example, a social security number field may have one of the following data names:

SOC-SEC-NUM
SS-NUM
SSN

Such abbreviations save time and space in coding specifications and programs. However, an abbreviation should not be carried to the point that it becomes difficult to tell the original full name of the data element. If the programming language to be used is not known when the data names are created, the names can be modified, if necessary, after the language has been determined.

Definition

The definition provides both a field format and a narrative description of the meaning of the variable. For example, QUANTITY-ON-HAND can be defined as the number of items in existing inventory, excluding any back-orders and any items committed to customers but not yet shipped. If a data element is to be expressed as a code, the code values and their meanings should be included in the definition. For example, a code for sex may be established as follows:

Code Value	Meaning
1	Male
2	Female

The format of a field is defined by indicating both the length of the field and whether the data to be stored in the field are numeric or alphanumeric. "9s" and "Xs" are used to represent numeric and alphanumeric characters, respectively. For numeric fields, the allowable sign values (i.e., + and/or −) should be indicated. If a decimal point is involved, its position should be defined.

Source

Data elements can come from one of three places: inputs, tables, or computations. Inputted data elements come directly into the system from an input data stream (e.g., a transaction file). Table data elements are looked up in predefined tables (e.g., tax-rate tables). Computed data elements are computed from other data elements in the system.

For a data element that is inputted, the input source is identified, and any input validation rules are defined. For example, a validation rule for the data field of a transaction may be stated as follows:

> *Validation Rule:*
> Date value cannot be later than today's date or more than 30 days earlier than today's date.

For a data element to be determined by a table lookup, the table values and the indexes used for the table lookup are defined. This can be accomplished by simply displaying the table. The example below illustrates how a SALARY table in which the table indexes are PAY-GRADE and PAY-STEP may be defined.

SALARY Table:

	PAY-STEP				
PAY-GRADE	1	2	3	4	5
1	500	532	568	612	688
2	645	720	785	822	876
3	832	912	986	1666	1188
4	1112	1186	1290	1380	1440
5	1360	1502	1612	1720	1816
6	1706	1820	1920	2012	2118

PAY-GRADE and PAY-STEP are data elements that must be defined elsewhere in the data element dictionary. The SALARY table is used by locating the point of intersection for a given set of values for PAY-GRADE and PAY-STEP. For example, a PAY-GRADE of 4 and a PAY-STEP of 3 determine a value of 1290 for SALARY.

Table values may be maintained external to a program (e.g., on cards or disk) and inputted during program execution. Alternatively, they may be coded as part of the program. If table values change frequently, it is preferable to maintain them external to the program so that they can be changed without modifying program code.

For a data element that is computed, the necessary computation is defined. This can be done in simple algebraic form with one minor variation. The equal sign (=) in computer processing means that the variable on the left side of the equation is "set equal to" the computation or variable on the right side of the equation. For example, if $A = 6$ and $B = 3$, the equation $C = A + B + 2$ results in C being set equal to 11, irrespective of C's value prior to the computation.

Each of the data elements used to compute the value for a computed data element must be defined elsewhere in the data element dictionary. For cross-reference purposes, care must be taken to insure consistent spelling of the name of each data element.

If a sequence of computations is needed to compute a data element, the complete sequence must be defined. For example, assume the data element to be computed is BALANCE-DUE. The sequence of computations shown at the top of the next page may be used:

Computation:

1. PAYMENT = CASH-DEPOSIT + CHECK-DEPOSIT
2. NEW-BALANCE = OLD-BALANCE − PAYMENT
3. INTEREST = NEW-BALANCE × INTEREST-RATE
4. BALANCE = NEW-BALANCE + INTEREST

There may be logical processes required for inputting or computing data elements. In a simple case, the logical process may be described in the data element dictionary. For example, consider the following logic associated with the computation of overtime:

IF HOURS WORKED IS GREATER THAN 40, DO COMPUTATION 2 OTHERWISE DO COMPUTATION 1.
1. GROSS-PAY = HOURS-WORKED × PAY-RATE
2. OVERTIME = HOURS-WORKED − 40
   GROSS-PAY = (40 × PAY-RATE) + (OVERTIME × (1.5 × PAY-RATE))

For complex logical relationships, decision tables (discussed in the following section) should be used. Decision tables are assigned numbers and/or names so they can be referenced from the data element dictionary as follows:

Computation:
Refer to Decision Table 1, OVERTIME-COMPUTATION, to determine which of the following computations to use:

1. Computation 1
2. Computation 2

.
.
.

n. Computation *n*

Where Used

A where-used list is created to keep track of each report and computation in which a data element is used. Such a list is needed for control purposes.

Consider a situation where the data element SALARY is printed on several reports and displays available throughout the organization. If the organization (or the government) decides distribution of this salary information should be restricted, the where-used list pinpoints the reports and displays that should be reviewed.

In another situation, the definition of a data element may be changed. For example, assume the data element QUANTITY-ON-HAND is redefined to include back-orders. In such a case, it is judicious to review all computations in which QUANTITY-ON-HAND is used to check for unanticipated effects on computations. Also, those persons receiving reports affected by this change can be identified and alerted to the new procedure.

Maintenance

The maintenance information about a data element describes the procedures for maintaining or updating the element. The concept of documenting maintenance is best illustrated by examples.

SEX-CODE:	Originated when the employee is first put on the payroll. No maintenance is required for this data element unless an error is detected.
BALANCE-DUE:	Computed during the monthly billing cycle for each customer record in the accounts receivable file.
SALARY:	Updated after each annual review unless a non-scheduled merit increase is granted.

Storage

Storage defines if and where (i.e., in what file) a data element is stored. Not all data elements are stored in a file. For example, in processing a weekly payroll for hourly employees, GROSS-PAY is computed by multiplying PAY-RATE by HOURS-WORKED. Since HOURS-WORKED is inputted each week and GROSS-PAY is computed each week, neither data element has to be stored in the payroll file. In contrast, PAY-RATE changes only when a raise is given. It is relatively stable from pay period to pay period. Consequently, PAY-RATE should be stored so that it is conveniently available for payroll processing. Storage for PAY-RATE is defined as follows:

Storage: Payroll Master File

Storage for GROSS-PAY and HOURS-WORKED is defined as follows:

Storage: Not Applicable (or N/A)

At the time the data element dictionary is established, decisions as to whether certain data elements are to be stored may not have been made. In such cases, the storage definitions may remain blank until the decisions are made.

Organization

There is no standard form for a data element dictionary. Since the amount of documentation needed for a data element varies, defining a data element is best accomplished on blank paper or index cards using the definition categories in an outline format. Figure 6-7 shows how a completed entry in the data element dictionary might appear.

For reference purposes, it is convenient to maintain the data element dictionary in alphabetical order. This may be difficult if

Figure 6-7. *Documentation for a Data Element*

Name	DEPARTMENT
Definition	The code of the department an employee works in. Format is 999 Code values are as follows: 101-Administration 201-Production 301-Sales 401-Engineering
Source	Input: New employee transaction Validation rule: Code value must be equal to a valid department code.
Where Used	Payroll Check Validation Error Report Department Payroll Report
Maintenance	Originated from new employee transaction when employee is hired. Updated whenever an employee changes departments.
Storage	Payroll Master File

unanticipated data elements are added frequently during design. However, if each data element is documented on a separate sheet of paper or index card, insertion (and deletion) of elements is accomplished easily. It may be possible to use an online system in setting up and maintaining the data element dictionary. This is a desirable enhancement but not a necessary one.

After all data elements are defined, they can be organized into one or more files. This is accomplished by clustering related data elements into groupings. For example, the data element dictionary for an order processing system contains data elements relative to both customers and inventory. This leads to the creation of a file for each of these groups.

Decision Tables

Decision tables are powerful, efficient tools for expressing complex logical relationships in an understandable manner. Decision tables may be used in conjunction with the data element dictionary and output definitions as follows:

1. They may be used in conjunction with the "source" section of the data element dictionary to define logic used for input validation or computations.
2. They may be used in conjunction with the "output selection logic" section of the output definition to define logic used for selecting records to be included in a report or a terminal display.

Terminology and Structure

A decision table is a matrix containing columns and rows that are used to define relationships. Figure 6-8 illustrates a decision table format.

There are three major components in a decision table: conditions, courses of action, and decision rules. *Conditions* are events or facts that determine the courses of action to be taken. *Courses of action* are processes or operations to be performed under certain conditions. *Decision rules* express the relationships between combinations of conditions and courses of action.

Decision rules are expressed by making condition entries and course-of-action entries in the matrix provided to the right of condition statements and course-of-action statements (see Figure 6-8). Possible decision-rule entries and their definitions follow:

1. Condition Entries
   Y = Yes, condition statement must apply.
   N = No, condition statement must not apply.
   – = Indifferent, condition statement is irrelevant to the decision and does not have to be checked.
2. Course of Action Entries
   X = Activate course of action
   . = Do not activate course of action.

Decision Table Example

Figure 6-9 shows a decision table that defines customer billing for the output selection logic of an output definition. Note that a double horizontal line is used to separate conditions from courses of action.

If a customer's balance is less than or equal to zero (decision rule 1, read vertically) no additional conditions have to be checked. A statement is not to be printed for the customer. If a customer's balance is greater than zero, but less than his or her credit limit (decision rule 2), no additional conditions have to be checked. A statement is to be printed for the customer. If the customer's balance exceeds his or her credit limit, then the credit rating is checked. If the credit rating is excellent (decision rule 3), a statement is to be printed for the customer. If the credit rating is not excellent (decision rule 4), a statement is to be printed and a credit warning message is to be added to it.

Constructing Decision Tables

The approach used to construct decision tables varies among systems analysts. The technique discussed below is based on the *progressive rule development* approach advocated by Keith London.[2] This technique is easy to understand, and it keeps the logic of the problem in sight at all times. The formal rules for constructing a decision table using progressive rule development are:

1. Conditions and courses of action should be listed in the sequence in which they are to be considered.
2. Consider a condition (relevant to the decision) in the positive case (Yes).
3. Consider, in the positive case, all other conditions (relevant to the decision) that must be considered before a course of action can be taken.

---

2. Keith R. London, *Decision Tables,* 1972.

DECISION TABLE FOR ________________ APPLICATION

DECISION TABLE NAME ________________ REFERENCE NO. ________

PREPARED BY ________________ DATE ________________

CONDITIONS/ COURSES OF ACTION	DECISION RULES																																																	
	01	02	03	04	05	06	07	08	09	10	11	12	13	14	15	16	17	18	19	20	21	22	23	24	25	26	27	28	29	30	31	32	33	34	35	36	37	38	39	40	41	42	43	44	45	46	47	48	49	50

Figure 6-8. *Layout Form for a Decision Table*

CONDITIONS/ COURSES OF ACTION	DECISION RULES											
	1	2	3	4	5	6	7	8	9	10	11	12
BALANCE-DUE ≤ 0?	Y	N	N	N								
BALANCE-DUE > 0 AND < CREDIT-LIMIT?	—	Y	N	N								
BALANCE-DUE > CREDIT LIMIT?	—	—	Y	Y								
CREDIT RATING EXCELLENT?	—	—	Y	N								
DO NOT PRINT STATEMENT	X	•	•	•								
PRINT STATEMENT	•	X	X	X								
ADD CREDIT WARNING TO STATEMENT	•	•	•	X								

Figure 6-9. *Completed Decision Table*

4. Use the indifference code ("—") for any conditions that are irrelevant to the decision rule under consideration.
5. Enter the conditions, courses of action, and their respective decision rule entries in the decision table.
6. Start a new decision rule by negating the last positive condition in the current decision rule (leave all other entries the same).
7. Repeat from step 3 onward until the table is complete.

The decision rules in Figure 6-9 were determined using progressive rule development. Review the rules in Figure 6-9 (sequentially, 1 through 4 in the context of progressive rule development).

Two additional examples of decision tables are provided in Figures 6-10 and 6-11. Note the use of progressive rule development in both examples.

Figure 6-11, a decision table for airline reservations, introduces a new concept. The condition "First class available?" is stated as the third and sixth conditions. This repetition is required to adhere to the rule of listing entries in the sequence in which they are considered. In decision rules 5 through 8, the passenger is requesting economy-class. Therefore, there is indifference about first-class availability until it is determined that no economy seats are available and an alternative is acceptable. At that point, the next condition to be considered is the availability of a first-class seat.

The repetition of condition statements or course-of-action statements is occasionally necessary to insure clarity when using progressive rule development. This type of presentation is helpful

CONDITIONS/ COURSES OF ACTION	DECISION RULES									
	1	2	3	4	5	6	7	8	9	10
Salaried?	Y	Y	N	N	N					
Hourly?	—	—	Y	Y	Y					
Hours worked < 40?	Y	N	Y	N	N					
Hours worked = 40?	—	—	—	Y	N					
Hours worked > 40?	—	—	—	—	Y					
Pay base salary	X	X	•	•	•					
Calculate hourly wage	•	•	X	X	X					
Calculate overtime	•	•	•	•	X					
Produce Absence Report	X	•	X	•	•					

Figure 6-10. *Decision Table for Payroll Calculations*

to computer programmers because the logic can be translated conveniently into program code in the same sequence as it is presented in the decision table.

CONDITIONS/ COURSES OF ACTION	DECISION RULES							
	1	2	3	4	5	6	7	8
First-class request?	Y	Y	Y	Y	N	N	N	N
Economy-class request?	—	—	—	—	Y	Y	Y	Y
First-class available?	Y	N	N	N	—	—	—	—
Economy available?	—	Y	Y	N	Y	N	N	N
Alternative acceptable?	—	Y	N	—	—	Y	Y	N
First-class available?	—	—	—	—	—	Y	N	—
Reduce first-class available	X	—	—	—	—	X	—	—
Issue first-class ticket	X	—	—	—	—	X	—	—
Reduce economy available	—	X	—	—	X	—	—	—
Issue economy ticket	—	X	—	—	X	—	—	—
Refer to alternate flight	—	—	X	X	—	—	X	X

Figure 6-11. *Decision Table for Airline Reservations*

Form Identification

Each decision table should be identified for documentational purposes, preferably in an area provided on the decision table form. The minimum points to be documented are illustrated by the following example:

Name of Application or System:	Payroll System
Decision Table Name:	Pay Calculations
Reference Number:	112
Prepared by:	Jill Smith
Date:	5-30-80

Systems Flowchart

The systems specifications are completed by preparing a systems flowchart to graphically convey the flow of processing. Similar flowcharts, as used to analyze an existing system, were discussed in Chapter 5. Figure 5-2 presented the various symbols used in constructing a systems flowchart. Figures 5-3 and 5-4 showed examples of flowcharts.

Constructing a Flowchart

Since the other systems design specifications describe the inputs, processes, files, and outputs of the system, the systems flowchart need only link these various systems components. Each input document or display, and each output document or display, on the flowchart should be labeled exactly as its respective input or output definition is labeled.

Data elements can be organized into one or more files, and the files can be given file names. Figure 6-12 illustrates file contents for both inventory and customer master files used in an order processing system. The computer programs necessary for the input, processing, and output can then be determined, given names, and charted into the system. Computer programs are required between the following components:

System Components	Example of Program Processing Function
Inputs and outputs	List input transactions
Inputs and files	Update file(s) with transactions
Files and outputs	Create reports using file(s)
Files and files	Merge two files into one file

INVENTORY MASTER FILE	*CUSTOMER MASTER FILE*
ITEM-NUMBER	CUSTOMER-NUMBER
DESCRIPTION	CUSTOMER-NAME
CLASS	CUSTOMER-ADDRESS
VENDOR-NUMBER	ZIP-CODE
ITEM-WEIGHT	PHONE-NUMBER
OUT-OF-STOCK-CODE	YEAR-ACCT-OPENED
TAX-CLASS	CREDIT-LIMIT
UNIT-OF-MEASURE-CODE	CREDIT-RATING
UNIT-OF-MEASURE	CREDIT-RATING-DATE
SALESPERSON-COMMISSION-RATE	SALESPERSON-NUMBER
SUGGESTED-RETAIL-PRICE	SERVICING-BRANCH-CODE
PRICE-CODE	DATE-OF-LAST-SALE
PRICE-A	CURRENT-BALANCE
QUANTITY-BREAK-A	
PRICE-B	
QUANTITY-BREAK-B	
PRICE-C	
QUANTITY-BREAK-C	

Figure 6-12. *File Contents for an Order Processing System*

Documentation of any required data preparation and report distribution steps completes the flowchart.

DESIGN REVIEW

Usually, several persons are involved and several weeks or months elapse in preparing systems specifications. Due to the number of people involved, the amount of time required, and the complexity inherent in developing systems specifications, there are frequently omissions, inconsistencies, and other errors that are not recognized until the various specifications are brought together.

An effective technique to highlight any problems with systems specification is a design review. A design review is a formal presentation of the systems specifications. It is a "guided tour" of the system by those who developed the specifications.

The audience for the design review may consist of interested users and/or managers and, of course, members of the systems design team. The design review is a learning experience for all involved.

Using the systems flowchart as the nucleus, a design review draws from the input and output specifications, data element dictionary, and decision tables to provide a detailed review of systems functions and capabilities. The design review is an excellent tool for uncovering errors and eliminating them. It provides positive motivation for those involved in preparing systems specifications and provides a measure of completeness before proceeding with systems development.

SUMMARY

Systems design is involved with both inclusion and structuring of the information system as it pertains to information, technology, and personnel. The systems analyst plays the lead role in the structuring of information to effectively and efficiently support the organization's information requirements.

The systems analyst, in conjunction with management, develops systems specifications. The following design documents are usually required:

- Output Definitions
- Input Definitions
- Data Element Dictionary
- Decision Tables
- Systems Flowchart

The output definitions describe printouts and terminal displays. The input definitions define the format of each data element coming into the system. The data element dictionary defines all data fields that are inputted, computed, stored, and reported. Decision tables illustrate complex logical relationships. The systems flowchart provides a graphical representation of the system.

These specifications are linked together such that the input, computations, logic, storage, and reporting of each data element can be traced forward or backward from its point of origin. The completed systems specifications provide sufficient documentation for the computer programming of the system.

EXERCISES

1. Show how to define a six-character, numeric data field that is to have a decimal point between the hundreds and tens positions. How many positions on an output definition does this field require? How would the field appear with a field value of 30? 488? 5069?
2. Construct an output definition field format for the following inventory part codes: AE91609, CG41113, and FR60021.
3. How is an item in an output definition linked to the data element dictionary?
4. Create an input definition format for the following input transaction:

   The first field is a nine-digit account number.
   The second field is a fifteen-character field for customer name.
   The third field is a six-digit date field in the format of month, day, year.
   The fourth field is a two-digit transaction code.
   The fifth field is a six-digit field called AMOUNT.
5. Where does a systems analyst define the computations and/or the reports in which a data element is used?
6. Why is it essential to assign unique names to data elements and always use their correct spellings?
7. What are the three possible sources of data elements in an information system?
8. Why is it important to document where data elements are used?
9. What determines whether a data element should be stored in a file?
10. Use decision tables to illustrate the logic for the following situations.
    a. ABC Inc. must deduct city tax from employee wages, depending upon where they live and work as follows:
       (1) Employees living in the city must pay the tax.
       (2) Employees living outside the city and working within the city must pay the tax.
       (3) Employees living and working outside the city do not pay the tax.
    b. Salespersons for XYZ Distributing have the possibility of receiving a monthly bonus check. To receive a bonus, a salesperson must achieve the following goals:
       (1) He or she must have sold in excess of ten units or sold orders to a minimum of two new customers.
       (2) He or she must have installed three units or sold an additional five units or sold an additional unit to a new customer.

(3) He or she must have turned in all expenses for the past months and have submitted reports on contacts of at least five new prospects.

SELECTED REFERENCES

London, Keith R. *Decision Tables.* Pennsauken, N.J.: Auerbach, 1972.

Lucas, Henry C. *The Analysis, Design, and Implementation of Information Systems.* New York: McGraw-Hill, 1976.

Montalbano, Michael. *Decision Tables.* Palo Alto: SRA, 1974.

National Cash Register Company. *Accurately Defined Systems.* Dayton, Ohio, 1968.

Nolan, Richard L. "Systems Analysis for Computer Based Information Systems Design," *Data Base,* Winter 1971, pp. 1–10.

Semprivivo, Phillip C. *Systems Analysis: Definition, Process, and Design.* Palo Alto: SRA, 1976.

Schussel, George. "The Role of the Data Dictionary," *Datamation,* June 1977.

Wetherbe, James C. "A Systems Specification Model for Instruction in Systems Analysis and Design," *Journal of Data Education,* July 1978.

Wetherbe, James C. "Development and Application of Industry-Based Cases in Systems Analysis and Design," *Journal of Data Education,* October 1978.

CASE STUDY

A Payroll System

Information Required from the System
Pay Statement
Paycheck
Error Listing Report
New Employee and Termination Report

Inputs to the System

Requirements to Complete Case

This case study applies the systems design and specifications concepts and techniques discussed in preceding chapters. The system to be designed is a payroll system. A payroll system is used because it is an information system common to all organizations.

The case is structured to be realistic, but it is necessarily abbreviated so that it can be completed in a short period of time. There are many ways the case can be expanded into a major, time-consuming project. To avoid doing this, you should stay within the requirements defined in the case.

INFORMATION REQUIRED FROM THE SYSTEM

The following basic reports are necessary for decision making and/or for documenting transactions (e.g., for auditing purposes).

Pay Statement (See Figure 1.)

1. Employee Number (EMP-NO)
   A six-character, numeric field with no zero suppression.
2. Employee Name (NAME)
   A twenty-character, alphabetic field.
3. Social Security Number (SOC-SEC-NO)
   A nine-character, numeric field with hyphens between the fourth and fifth positions from the right and the sixth and seventh positions from the right.
4. Hours Worked (HRS)
   A four-character, numeric field. It is to have a decimal point between the tens and units positions and zero suppression to the decimal point. This is the total hours worked for the week.
5. Department (DEPT)
   A five-character, numeric field with zero suppression in the two high-order positions. The employee will not work in more than one department per week. If there is a departmental transfer within the system, it will be picked up the following week through a change record. The department numbers range from 100 to 99999.
6. Date Check Is Written (DATE)
   A six-character, numeric field with month, day, year format. The month and day fields should have the possibility of zero suppression in their high-order positions. There should be slashes (/) between the month and day, and the day and year. This is today's date (actual date). It will be available within the computer.
7. Check Number (CK-NO)
   A five-character, numeric field with zero suppression up to but not including the units position. A consecutive check number is to be printed on each check. The range is from 1 to 99999. When the upper limit is reached, this field should be reset to 1.
8. Earnings
   The fields under this heading represent all possible ways to earn income. All fields are to have decimal points between the hundreds and the tens positions.
9. Regular Earnings (REG)
   A seven-character, numeric field with zero suppression up to the decimal point. Regular earnings is basic pay less any overtime income, which is to be itemized separately on the next line of print. A salaried employee's wages are considered

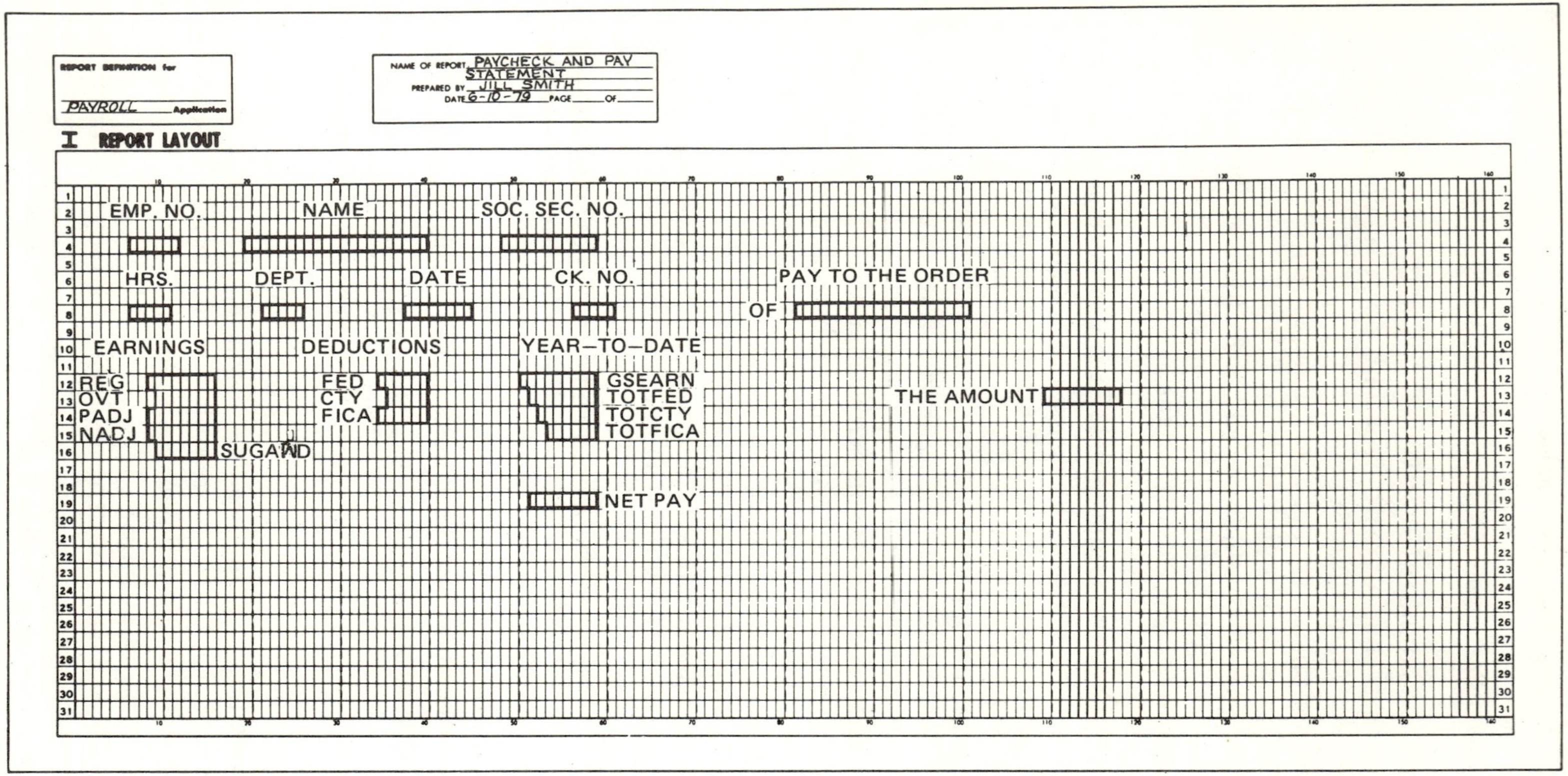

Figure 1. *Format of Paycheck and Pay Statement*

regular earnings. Regular earnings for an hourly worker are hours (≤ 40) times the rate. The employee will have only one rate per week.

10. Overtime Earnings (OVT)
A six-character, numeric field with zero suppression up to the decimal point. Overtime earnings can be received only by an hourly employee. Overtime is based on hours over 40. All hours worked over 40 are considered overtime income hours. The overtime rate is 1.5 of the regular rate for each hour of overtime.

11. Positive Pay Adjustment (PADJ)
A seven-character, numeric field with zero suppression up to the decimal point. There can be only one positive pay adjustment per week per employee. Any employee may have a positive pay adjustment.

12. Negative Pay Adjustment (NADJ)
A seven-character, numeric field with zero suppression up to the decimal point. There can be only one negative pay adjustment per week per employee. Any employee may have a negative pay adjustment. It will never exceed the net pay for the week. There will be approximately ten to twenty of these per pay period. (An employee can receive both a positive and a negative pay adjustment on the same check.)

13. Suggestion Award (SUGAWD)
A six-character, numeric field with zero suppression up to the hundreds position (dollar position). All employees are eligible for a suggestion award. The minimum award is $1.00, and the maximum award is $6000.00. An individual may have more than one award per week. If he or she has more than one, the field on the check is to be a summation (total) of the awards.
NOTE: The identifying captions REG, OVT, PADJ, FED, GSEARN, and the like, are not preprinted. If no data entry is to be made in a field, do not print the caption. Do not leave a blank line of print; shift the next entry up in its place. If no entries remain for a column, proceed to next print step. Except for these captions under Earnings, Deductions, and Year-to-Date, the rest of the check is preprinted.

14. Deductions
The following three fields represent the possible deductions. Each field is to have a decimal point between the hundreds and the tens positions, and each is to have zero suppression to the decimal point.

15. Federal Withholding Tax (FED)
A five-character, numeric field. This tax is to be deducted for

all employees. The tax is a percentage of gross income for the week. This percentage is given in Table 1. It is based on the projected yearly earnings and the number of dependents claimed by the employee. (Projected yearly earnings are the total weekly earnings times 52.)

16. City Tax (CTY)
A four-character, numeric field. This tax is 1 percent of the employee's total earnings for the week.

17. Federal Insurance Contribution Act (FICA)
A five-character, numeric field. FICA (social security tax) applies to all employees. It is 6.05 percent of the gross income up to $17,700 for the year. (This formula is subject to change.) When the individual's earnings reach the $17,700 limit, this deduction is to cease for the rest of the current calendar year. NOTE: Do not overdeduct.

18. Year-to-Date
The following fields are to be the current status of these categories as of this pay period. All fields should have decimal points between the hundreds and the tens positions with zero suppression to the decimal point.

19. Gross Earnings (GSEARN)
An eight-character, numeric field that is a total of all gross earnings of each week (year-to-date), including the current week.

Table 1. *Federal Withholding Tax Table Deduction Percentage*

Projected Yearly Earnings	*DEPENDENTS*						
	0	*1*	*2*	*3*	*4*	*5*	*6 & up*
Less than $6,000	0	0	0	0	0	0	0
$6,001 — $8,000	16	18	16	14	13	11	9
$8,001 — $12,000	18	16	14	12	11	18	9
$12,001 — $15,000	22	20	18	16	15	14	13
$15,001 — $20,000	24	22	20	19	18	17	16
$20,001 — $35,000	28	26	24	23	22	21	20
$35,001 — $60,000	32	30	29	28	27	26	25
Greater than $60,000	36	34	32	30	29	28	27

20. Total Federal Withholding Tax (TOTFED)
    A seven-character, numeric field that is the total of all withholdings of this weeks tax, year-to-date.
21. Total City Tax (TOTCTY)
    A seven-character, numeric field that is a total of all deductions made each week (year-to-date) for city tax. The same rules apply here as when it is deducted.
22. Total FICA (TOTFICA)
    A five-character, numeric field that is an accumulation of this deduction year-to-date. It will become a fixed figure when the upper limit is reached.
23. Net Pay (NET PAY)
    A seven-character numeric field with a decimal point between the hundreds and the tens positions with zero suppression up to the decimal point. Net pay is the total earnings for the week minus the total deductions for the week. This amount will never be negative.

Paycheck (See Figure 1.)

1. Employee Name
   The name field (Pay to the Order of) is a twenty-character, alphabetic field. This name is to appear on the statement (STUB).
2. Check Amount
   A seven-character, numeric field with a floating dollar sign to the decimal point. The dollar sign is to be displayed in the first character position to the left of the first significant digit. A decimal point should appear between the hundreds and the tens positions.

Error Listing Report

This report is to display all input records having a field or fields that failed to satisfy validation rules. Place an asterisk under each field in error. The format of this report is not fixed. Create your own. Make it neat and understandable.

New Employee and Termination Report (See Figure 2.)

This report is a combination of two reports. All hirings and terminations for the week are to be listed on this report. If the person is a

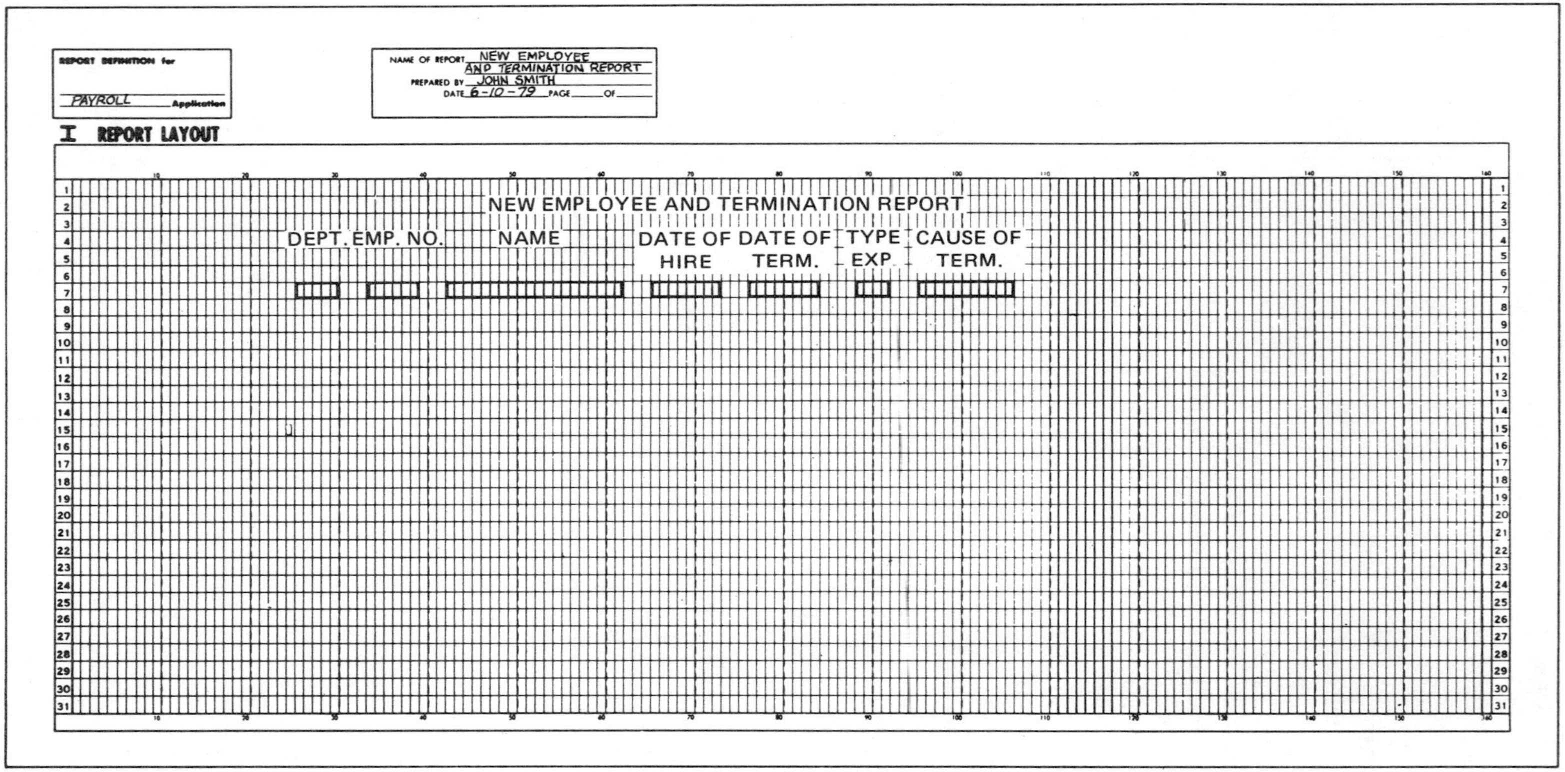

Figure 2. *Format of New Employee and Termination Report*

new employee, all fields except Date of Termination and Cause of Termination are to be printed. If the person is leaving, all fields are to be printed.

1. Department (DEPT)
   A five-character, numeric field with zero suppression in the first two high-order positions.
2. Employee Number (EMP-NO)
   A six-character, numeric field with no zero suppression.
3. Employee Name (NAME)
   A twenty-character, alphabetic field.
4. Date of Hire
   A six-character, numeric field with month, day, year format. The month and day may have zero suppression in their high-order positions. Slashes (/) are to be inserted between the month and day and between the day and year.
5. Date of Termination
   This field has the same format as Date of Hire.
6. Type Employee (TYPE-EMP)
   This field is to identify the type of worker—hourly or salaried. If an hourly worker, the letters HRLY are to appear in this field. If the employee is salaried, the letters SAL are to appear in this field.
7. Cause of Termination
   There are three possible reasons for termination of pay:
   a. Transferred out of the realm of this payroll system. If an individual is transferred, the word TRANSFERRED is to appear in this column. (This is not a departmental transfer within the payroll system. The individual is being transferred out of this payroll system to another, yet within the same corporation.)
   b. Terminated by the employer. If the employer terminates the individual, the word TERMINATED is to appear in this column.
   c. Terminated voluntarily. If an employee leaves on his or her own accord, the word VOLUNTARY is to appear in this column.

   Only one of these words can appear per line of print.

INPUTS TO THE SYSTEM

The inputs to the system are entered once each week. The formats of certain fields are identified and described below. These fields alone are not adequate to provide a complete solution to the problem. You must decide what other fields are needed to solve this payroll problem.

The following fields will always be present for a current employee:

1. Employee Number
   A six-character, numeric field.
2. Date of Reporting
   A six-character, numeric field with month, day, year format. This date cannot be greater than today's date or less than fourteen days prior to today. It is merely the date that the information is reported into the system, e.g., week-ending date.
3. Hours Worked for the Week
   A four-character, numeric field. This field cannot exceed the maximum number of hours in the week. The right-most position is tenths of hours. When used in computing, the field should be rounded to the nearest hour.

The following fields will always be present for a new employee:

1. Employee Number
   A six-character, numeric field.
2. Employee Name
   A twenty-character, alphanumeric field.
3. Birth Date
   A six-character, numeric field with month, day, year format.
4. Date of Hire
   A six-character, numeric field with month, day, year format. This date cannot exceed today's date or be less than fourteen days prior to today.
5. Hourly or Salaried Code
   A code that signifies whether an employee is hourly or salaried.
6. Rate or Wage
   A seven-character, numeric field. If the individual is a salaried employee, the field contains his or her weekly wage. The right-most two characters of the wage are cents positions. If the employee is on an hourly rate, the rate is in this field. The range is $2.25 to $15.00 per hour. The right-most two characters of the rate are cents positions. The employee cannot have

more than one rate per week. Both wage and rate cannot appear in a given record.

REQUIREMENTS TO COMPLETE CASE: Complete or design all output and inputs to the system; create a data element dictionary; use decision tables as necessary; and develop a systems flowchart. Stay within the requirements of the problem.

7 Systems Design and Selection of Technology and Personnel

In the preceding two chapters of this section, the information to be included in an information system and the structuring of that information were discussed. This chapter discusses concepts and techniques for determining and evaluating technology and personnel required to support an information system. Determining and evaluating technology and personnel is a process in which both management and systems analysts generally play an active role. This chapter fits into the information systems framework and relates to the preceding two chapters as follows:

	Inclusion	*Structure*
INFORMATION	Chapter 5	Chapter 6
TECHNOLOGY	Chapter 7	
PERSONNEL	Chapter 7	

The chapter is organized into three major sections: Inclusion, Technology, and Personnel. The first section discusses general concepts of selection applicable to both technology and personnel. The next two sections address unique selection issues as they pertain to technology and personnel.

INCLUSION

Systems Approach

The systems approach discussed in Chapter 2 provides a framework for determining what technology and personnel to include in a system. To review, this framework involves the following steps:

1. Establishing systems objectives.
2. Determining and selecting entities or components possessing the necessary attributes to accomplish the objectives.
3. Structuring or relating the attributes of the entities included in the system such that the objectives are obtained.

The first two steps are relevant to this chapter; the third step is of primary interest in the next chapter.

The systems objectives are based upon the capabilities required to support an organization's information requirements. For example, an objective might be to provide a two-second response time to inquiries about the balances of customers' savings accounts. Given this objective, alternative combinations of both technology and personnel capable of achieving it should be determined. The most cost-effective combination should be selected for inclusion in the information system. Personnel, in this context, pertains to both development and operational personnel. Development personnel are primarily involved with creating the application programs needed in the system. Operational personnel are primarily involved with using the completed system to provide information.

Decision Making

The selection of technology and personnel for inclusion in an information system is a decision-making process. In Chapter 5 (see Table 5-2), decision making was defined as a process involving intelligence, design, and choice.[1] It is helpful to consider inclusion decisions in the context of intelligence, design, and choice.

Intelligence

During the intelligence phase, those persons involved in the decision-making process learn about technology and personnel available to support the desired information system. Selection and evaluation criteria should be based upon the attributes required of the technology and the personnel. For example, important attributes for a computer programmer may include COBOL expertise, accounting background, and online experience. The importance of these attributes should, of course, be related to the technology to be used. In this example, the technology is likely to include an online computer system with a COBOL compiler that is to be used for an accounting application.

Design

With an adequate level of intelligence about both the problem and the possible means (technology and personnel) to address it, various combinations of technology and personnel should be considered. These combinations may be alternative means to achieving the solutions defined during systems design for information (i.e., Chapter 6). For example, rather than hire COBOL programmers, an alternative may be to contract with another company to write the required programs. Selection criteria may be further defined as a better understanding of alternatives is determined.

The best alternative can be selected only if that alternative is recognized and available for consideration. It is important not to look just for an alternative that will work. The decision maker should attempt to identify all reasonable alternatives that deserve consideration.

Choice

Choice involves selecting the alternative that best satisfies or approximates the selection criteria established during the intelligence and design phases. A combination of both economic and subjective analyses is usually required. For example, the difference in cost between two pieces of equipment can be analyzed, but there are also subjective considerations such as the reputations of

---

1. Herbert Simon, *The New Science of Management Decision*, 1960, p. 54.

the equipment suppliers and the capabilities of their technicians. Such subjective considerations can usually be quantified; however, care must be taken to remember that the numbers do represent subjectivity.

TECHNOLOGY

The systems approach of considering objectives and technology with attributes that can contribute to achieving objectives is an important dimension to selecting technology. In a technologically dynamic field such as computing, it is unfortunately common for those responsible for technology decisions to continue to use hardware and software with which they are familiar, rather than thoroughly investigate and consider other technological alternatives. To illustrate, the use of minicomputers was a cost-effective alternative for online processing several years before most organizations considered implementing them as an alternative to large centralized computers. Even computer professionals are occasionally reluctant to change to technologies that differ from those current when they received their training. In other words, they tend to invest their efforts in structuring technology with which they are familiar to accomplish something new, rather than back up to consider whether that technology is still the most effective and efficient.

The reluctance of a computer professional to capitalize on new technologies is not dissimilar to that of an accountant who resists changing from a manual accounting system to a computerized one. No amount of structuring of the manual system can provide the performance that is possible with computer technology. There exists somewhat of a paradox when computer professionals, who have brought so much change on others, are reluctant to endure change themselves.

One of the greatest contributions of the systems approach to problem solving is that inclusion is considered prior to proceeding with structuring. It is far better to find the technology that can best perform the task required than to fit the task to a more familiar but less suitable technology.

Intelligence

To properly consider technology alternatives, the decision maker(s) responsible for technology selection must know about the "state-of-the-art" hardware and software technology. Research into existing technology is, therefore, required. The primary sources of intelligence about technology are:

- In-house expertise
- Vendor representatives
- Consultants
- Literature
- Other organizations

In-House Expertise

Computer professionals within the organization are a logical starting point when aiming to acquire information about technology that may be included in an information system. In very simple or obvious cases, they may provide the only expertise required. For example, if an organization is implementing a new personnel information system in conjunction with an existing payroll information system, simply using the existing computing equipment may be a clear-cut decision. However, for more complex and sophisticated situations, such as redesigning several major information systems to use online technology, additional expertise may be desirable.

A limitation of using in-house staff is that they are not likely to be aware of the latest technological developments available from all hardware and software suppliers. They may also be biased toward hardware and software with which they are familiar.

Vendor Representatives

Representatives from the various manufacturers of hardware and/or software (e.g., IBM, Burroughs, NCR, Honeywell, Univac, CDC, etc.) are a major source of information on technology. Vendor representatives can be contacted and requested to give presentations and demonstrations of their company's available or forthcoming hardware and/or software relevant to the information system under consideration.

The major drawback of vendor representatives as an information source is that they tend to be biased toward the hardware and/or software they are marketing. It is important to acquire information from several different vendor representatives to insure an objective perspective of alternative technologies.

Consultants

Consultants can be contracted to provide information on technology and to make recommendations as to selection. Their insight can be helpful because they generally have a wide variety of experience with different organizations and technologies. This additional perspective can be highly beneficial.

On the negative side, consultants cost money, are subject to hardware/software biases, as is any computer professional, and only make recommendations. They will not be around to live with the results of their recommendations.

Literature

The various professional and trade journals unique to the computing industry are an excellent source of information on technology as well as on other aspects of computing and information systems. It is worthwhile to subscribe to several of these publications. Some of the common publications are:

1. *Communications of the Association of Computing Machinery*
2. *Computer Decisions*
3. *Computer World*
4. *Computing Reviews*
5. *Data Communications*
6. *Data Management*
7. *Datamation*
8. *Infosystems*
9. *Mini-Micro Systems*
10. *MIS Quarterly*

Two other very useful sources of information on technology are available through industry research services provided by DataPro and Auerbach. Both companies market subscription services that provide information on hardware and software supplied by different companies. These sources can be used to determine vendors that should be contacted for additional information.

The shortcomings of literature as a source of intelligence are:

1. Articles in publications are generally a result of an author's (or authors') decision to make a contribution in a certain area. Therefore, there may or may not be articles that specifically discuss technology pertinent to the information system under consideration. The articles that are available may contain author biases.
2. Hardware/software information services, though extremely comprehensive, are, of necessity, generalized. They do not

specifically discuss hardware or software as it applies to a particular organization's problems.

Other Organizations

Other organizations that have developed or are developing similar computer-based information systems can provide considerable insight into the advantages and disadvantages of technology. A common way of locating such organizations is through vendor representatives. It is quite appropriate to request a list of organizations to which the vendor has sold similar hardware and/or software. In situations where there is some question as to the capability of a vendor's hardware and/or software, visiting with some of their other customers is highly desirable.

The disadvantages of information based on visits to other installations are:

1. Other installations are not identical so care must be taken not to inaccurately generalize.
2. The reasons for another organization's technology decisions may no longer apply due to new technological developments that provide better alternatives.
3. If the other organization is a competitor, the organization may not be willing to share all or any of the information about its environment.

General Selection Criteria

The intelligence phase provides a general understanding of what information systems capabilities are possible with existing technology. Determination can be made of the technological and economical feasibility of supporting the information system as desired. For example, in an online retail information system, are there combinations of retail terminals and computers that can perform the necessary processing functions at a price that is realistic for the organization? If the technology does not yet exist or is cost prohibitive, it may be necessary to compromise the desired information system to strive for one that is realistic.

Such intelligence about current technology and its cost allows the organization to establish general selection criteria for the type of system desired. Examples of such criteria are whether batch or online processing is to be used, a general idea of processor size and speed, storage capacity needed, and, of course, a price range. Preliminary criteria as to the type of system required provide the framework for the design phase of the decision-making process.

Design

During the design phase, the information about technology is structured into specific alternatives. The objective of the design phase is to design the most economical alternative for properly supporting the information system. There is generally a wide variety of alternatives. It is easy and also tempting to get overly involved with the rather exciting technological elegance of them. For example, it is interesting to compare processor instruction sets and speeds, disk access techniques, and operating software architecture. But alternatives should be compared in terms of *what they do,* not *what they are.*[2] This perspective can best be illustrated by analogy. In purchasing a television set, the type of technology used to produce a quality picture, insure reliability, and so forth, may be interesting. But the selection of a television set should be based on performance, or what the television does, not purely on technology. The value of technology is based on the results it produces, not in how interesting or complex it is.

RFQ and RFP

The end result to be accomplished during the design phase is the creation of systems specifications that can be used to request hardware and/or software proposals from vendors. These specifications are generally expressed in either a *Request for Quotation* (RFQ) or a *Request for Proposal* (RFP).

An RFQ is used when the hardware and/or software to be procured is somewhat predetermined. This usually applies in situations where an organization has existing computing facilities that need only be expanded to accommodate a new information system. In such cases, an additional disk drive or additional computer memory may be all that is required. Therefore, an RFQ that specifies the exact hardware and/or software required is prepared. The vendor(s) respond with specific prices. Figure 7-1 illustrates an RFQ for CRT terminals. Some technical terms in Figure 7-1 may be unfamiliar to the reader. Familiarity with the terms pertinent to an RFQ is acquired during the intelligence phase of a "real" selection exercise. For the purposes of this discussion, such familiarity is not required.

An RFP is used for more complex procurement activities. Generally, there is uncertainty as to the best hardware and/or

2. Erik M. Timmreck, "Computer Selection Methodology," *Computing Surveys,* Dec 1973, p. 201.

REQUEST FOR QUOTATION

Six CRT Terminals

Specifications:

1. Sixty-four Alphanumeric ASCII Characters, Upper and Lower Case
2. Cursor Addressing
3. Dual-Intensity Display Mode
4. 1920-Character (24 X 80) Screen
5. RS-232 Interface
6. 300/2400 Baud
7. Serial Printer Interface (include RS-232 Cable)
8. Half-Full Duplex Mode
9. Acoustic Coupled or Hardwire Connections
10. RS-232 Ten-Foot Cable
11. Ten-Key Numeric Pad

Figure 7-1. *An RFQ for CRT Terminals*

software to procure. Such situations occur when an organization is installing its first computer system or replacing all or most of an existing one. An RFP is less restrictive than an RFQ in that an RFP indicates the functional capabilities required but allows latitude to vendors in configuring and proposing different hardware/software alternatives. In fact, the RFP approach allows vendors to participate in the design of alternatives to meet an organization's requirements. For example, one vendor may propose both centralized and distributed technology for an online system. The

```
                    RFP
                    FOR
             ACME MANUFACTURING

   I.  INTRODUCTION

       A.  Purpose of Request for Proposal
       B.  Summary of Requirements
       C.  Background Information

  II.  GENERAL INFORMATION AND REQUIREMENTS

       A.  Schedule of Selection Process
       B.  Proposal Deadline
       C.  Vendor Contact with ACME Manufacturing
       D.  Alternative Proposals
       E.  Proposal Evaluation Criteria
       F.  Caveats

 III.  APPLICATIONS TO BE SUPPORTED

       A.  Production Scheduling
       B.  Material Requirements
       C.  Financial Systems

  IV.  SPECIFICATIONS

       A.  Representative Hardware Configuration
       B.  Representative Hardware Listing
       C.  Attributes and Support Requirements

   V.  CONTRACTUAL EXPECTATIONS

  VI.  EVALUATION OF PROPOSALS

       A.  General Approach
       B.  Hard-Dollar Evaluation
       C.  Soft-Dollar Evaluation

 VII.  SPECIFICATIONS FOR VENDOR PROPOSALS

       A.  Vendor Reply Format
       B.  Vendor Cost Schedule
```

Figure 7-2. *An Outline for an RFP*

organization can take advantage of the technological creativity available from a wide variety of computer professionals working for different hardware and/or software suppliers.

A detailed discussion of RFP's is beyond the scope of this text. Figure 7-2 summarizes the contents common to an RFP, in outline form.

Mandatory and Desirable Specifications

Specifications for RFQ's and RFP's are generally expressed in terms of mandatory and desirable specifications.

Mandatory specifications indicate hardware and software features and capabilities that must be available within an alternative to be considered. For example, a mandatory requirement for an online system might be the ability to concurrently support at least thirty terminals with an average response time of less than two seconds.

Desirable specifications indicate hardware and software capabilities that are of value but are not essential to the operation of the system. Desirable attributes pertain to enhancements to mandatory specifications or introduce new features or capabilities not discussed under mandatory specifications. For example, it may be desirable, but not necessary, for an online system to support more than thirty terminals.

Care must be taken in specifying both mandatory and desirable specifications. The writer must not use specifications that are too "vendor oriented" or unfairly restrictive. For example, it would be unfairly restrictive to specify that disk access time must be exactly 30 milliseconds if it were known that only one vendor's disk access time was exactly that speed. To avoid such unfairness, specifications should be expressed in terms of minimal acceptable levels. An example is:

> Disk access time must not exceed 40 milliseconds.

Hard and Soft Dollars

The criteria by which RFQ's and RFP's will be evaluated should be defined in the systems specifications. Obviously, a major selection criterion is the cost, or "hard dollars," required to purchase the hardware and/or software. Any response to an RFQ or RFP that meets all mandatory specifications should be given a "hard-dollar" evaluation. Those that do not meet the mandatory specifications should be considered nonresponsive to the specifications and eliminated from further consideration.

Systems specifications that are desirable may be assigned "soft-dollar" values. A "soft-dollar" value is a subjective assessment of the economic worth of a desirable feature or capability of a hardware or software component. "Soft dollars" are a mechanism to award vendors for features or capabilities provided above and beyond mandatory requirements. For example, vendors who can deliver hardware in advance of the mandatory delivery date might be given a "soft-dollar" credit against the "hard-dollar" cost of their equipment.

Vendors should be notified in advance as to what "soft-dollar" evaluations will be made of their quotations or proposals.

Providing this information allows vendors to assess the importance and, therefore, the need to address desirable specifications.

Figure 7-3 illustrates how mandatory and desirable systems specifications can be expressed. Note that vendors may be awarded different amounts of "soft dollars," depending on the levels at which they satisfy a desirable specification.

MANDATORY SPECIFICATIONS

I. Main Storage (Memory)

A. System must have minimum storage capacity of 4 million bytes.

B. Fetch cycle for one word should not exceed 1 microsecond.

C. Store cycle for one word should not exceed 2 microseconds.

D. Memory should exhibit physical modularity to enable functional isolation for trouble-shooting and fault detection.

E. Memory should permit the use of parity bits for error detection.

DESIRABLE SPECIFICATIONS

I. Delivery Date

Vendors will be awarded "soft-dollar" credits toward the cost of their proposals for bettering the mandatory delivery date. The schedule of credits follows:

Delivery Date	Credit
May 1981	$60,000
June 1981	40,000
July 1981	20,000
August 1981	-0-

Figure 7-3. *Mandatory and Desirable RFP Specifications*

Choice

The responses to RFQ's and RFP's provide the information required to proceed with choosing an alternative. The objective is to select the most cost-effective alternative.

Cost Analysis

In the simplest case, an RFQ or RFP may contain only mandatory requirements. In such a case, all that is necessary is a "hard-dollar" evaluation. The RFQ or RFP with the lowest cost is awarded the order.

In cases where desirable specifications are used, both "hard-dollar" and "soft-dollar" evaluations are required. The "hard-dollar" evaluation is conducted as in the preceding paragraph to determine the *actual cost* of each alternative. The "soft-dollar" evaluation is then conducted as follows:

1. For each desirable attribute provided by a vendor, the associated "soft-dollar" value is *deducted* from the vendor's "hard-dollar" cost.
2. The vendor with the lowest cost after "soft-dollar" adjustments is awarded the bid. If the system is to be rented or leased (rather than purchased), "soft-dollar" deductions are prorated over the anticipated life of the rental agreement. For example, if a "soft-dollar" adjustment of $60,000 is prorated for a system renting for $20,000 a month for five years, the adjusted price is $19,000 a month.

In some cases, desirable specifications may have "hard-dollar" costs. For example, additional hardware may be required to support additional (desirable) terminals. This can be adjusted for by conducting separate "hard-dollar" evaluations for vendors capable of supplying the particular feature or capability.

Proposal Validation

A vendor's response to a customer's proposal is a written document describing a system or systems that the vendor considers responsive to the customer's requirements. However, in an attempt to provide an attractive price, a vendor may propose a system that is inadequate; it may not be able to properly process the workload. Therefore, the customer should check whether or not each proposed system is indeed capable of handling the workload before a final choice is made.

The most accepted and comprehensive means of validating a proposal is the use of a benchmark.[3] A *benchmark* is a point of reference from which performance measurements of hardware and software can be made. Benchmarks are most commonly conducted by running computer programs and/or mixes of computer programs that are the actual work, or are representative of the actual work, to be performed on the new system. By measuring the times required to process such workloads, projections can be made as to the ability of the proposed system to process the total workload.

Benchmarks can be used to validate and compare the performance of various systems or just to validate the performance of the most economical proposal. Comparative benchmarks are advantageous in situations where the costs of different proposals are very close. In such cases, a slight performance advantage may be the deciding factor. If one proposal is substantially more economical than the others, a single benchmark may be conducted. If the most economical system performs satisfactorily, it may be selected. Otherwise, the next lowest proposal should be benchmarked. The benchmarking is continued until a system that performs satisfactorily is identified. Because each benchmark generally requires time and effort on the parts of both the customer and a vendor, the single benchmark approach should be used when appropriate.

PERSONNEL

As discussed at the beginning of this chapter, both development and operational personnel are required to support an information system. The need to make personnel selection decisions is invoked by the need to replace existing personnel (i.e., personnel who have been promoted, transferred, or terminated, or have resigned) or to expand the number of personnel.

The systems analyst is seldom involved in the selection of managers and clerical personnel to use and operate the system. However, systems analysts are often called upon to make recommendations about, and/or to participate in, the training of operational personnel.

The area of personnel selection in which the systems analyst usually participates is the selection of data processing

---

3. Eric M. Timmreck, "Performance Measurement: Vendor Specifications and Benchmarks," *The Information Systems Handbook,* 1975, pp. 365–367.

personnel such as programmers, other systems analysts, and managers for such groups. Therefore, the following discussion of personnel selection is oriented toward the selection of personnel who work directly in the data processing or information systems area. The framework is applicable to the selection of personnel for other positions as well.

Intelligence

In personnel selection, intelligence pertains to determining the capabilities required of personnel who are to participate in the development and maintenance of information systems. These capabilities requirements are based upon the characteristics of the information processed and the technology used to process it. For example, if an information system involves online technology for production scheduling, it is desirable to select personnel with related training and capabilities.

A careful review of the information systems and their respective technology provide the basic framework for selection criteria. Such criteria can be further defined during the design phase.

Design

The recruitment and selection of qualified personnel has been a difficult task for most organizations. The number of computer specialists available has not been sufficient to meet the demand for them. Title VII of the Civil Rights Act of 1964 and the guidelines set up within affirmative action programs have further complicated the task of personnel selection from a procedural perspective.

Now management's task is two-fold. First, a good personnel selection must be made. Second, the selection of one candidate over others must be justifiable if audited by an affirmative action officer or the Equal Employment Opportunity Commission (EEOC). Such justification can be particularly difficult if, as a result of national advertising for a position, several hundred applicants respond.

Detailed Descriptions

The common approach to personnel selection and affirmative action issues is to insure that job descriptions and specifications for positions are detailed and current. Well-defined position requirements can substantiate the selection of one candidate over others.

The use of position requirements has been fairly successful, particularly when selecting personnel for well-defined and relatively stable positions—for example, keypunch operators, control clerks, and computer operators. For more technologically diversified positions in programming, systems analysis, and management, job descriptions often lack the precision needed to serve as selection criteria and/or the flexibility needed for satisfactory selection of such personnel.

For example, in filling a programmer position, management may be looking for an individual with extensive payroll experience, familiarity with a certain programming language, and experience with a particular teleprocessing monitor on a certain hardware configuration using a certain operating system. In filling another programmer position in the same company, management may be interested in an individual with order-entry experience, the ability to utilize distributed processing with a specific minicomputer, and experience in using a programming language with specific intelligent terminals. A single programmer job description is inadequate for both.

Certainly, job descriptions can play a meaningful role in an organization. However, unless many job descriptions are maintained for many job classes (an unwieldy and continual task), job descriptions do not provide a comprehensive basis for personnel selection. The job descriptions for most data processing or information systems positions tend to be quite general and often fail to articulate the specifics required. Further, in most cases they do not provide sufficient documentation to support a given personnel selection, from an affirmative action or EEOC perspective, if necessary.

Selection Technique

As an alternative to using job descriptions as selection criteria, a quasi-quantitative selection technique has been developed. It provides both a clear definition of the selection criteria and comprehensive documentation for EEOC audits.[4] The six steps used in the technique are explained below. They provide a framework for the design phase of the decision-making process.

4. James C. Wetherbe and V. Thomas Dock, "Breaking the Description Dilemma: Personnel Selection by Group Analysis," *Data Management*, Dec 1976, pp. 16–18.

1. Formulate Selection Committee
   The selection committee should be a group of individuals competent in evaluating the candidates for a given position, preferably with a vested interest in the successful filling of the position. An essential committee member and likely chairperson is the immediate supervisor of the position to be filled. Other potential members of the committee include members of the candidates' potential peer group, individuals in subordinate positions to the position being filled, and members of departments who use services of the data processing or information systems department. A diversified committee is apt to follow a rigorous evaluation process. Also, a new employee will likely gain more acceptance from the organization.
2. Attribute Determination
   The committee should collectively identify attributes that are meaningful to the selection of a candidate for the position. Attributes should be categorized—education, work experience, test scores (if applicable), management background (when relevant), and so on. A sample profile of attributes for a management position is given in Figure 7-4.
3. Attribute Weighting
   At this point in the process, the committee assigns a weight to the value of each attribute as it relates to the position under consideration. One way to do so is by employing a scale of 1 through 10. This is a critical step; it prevents overemphasizing or overlooking an attribute or a category for a particular posi-

Attribute	Weight
Education	
EDP Training	8
Bachelors	10
Masters	3
	21
Work Experience	
Financial Applications	8
COBOL	8
Assembler	5
Online	6
General EDP	8
	35
Management Experience	
Supervisory Responsibilities	8
Professional Training	4
Communication	8
	20

Figure 7-4. *Attributes and Weights for the Position of Manager of Programming for a Bank*

tion. For example, technical attributes are weighted higher for a programmer than for a manager. Conversely, communication skills may be weighted higher for a manager than for a programmer. Sample weights for attributes are also shown in Figure 7-4.

4. Score Applicants

The scoring of applicants is accomplished by having each committee member independently rank each applicant by attribute on a scale of 1 through 5. Alternatively, the scoring of applicants may be performed collectively by the committee. However, on occasion, committees of this nature can be unduly influenced by a dominant committee member.

The scoring of applicants is particularly significant. This step provides the means for determining and also justifying the selection of one applicant over another. Therefore, it is important that the scoring criteria be fair and justifiable. For example, for most professional data processing or information systems positions, it would be appropriate to score an applicant with a degree in computer science and/or information systems higher than an applicant with a degree in general business. However, the applicant with a degree in general business would likely be scored higher than an applicant with a degree in, say, history.

If anonymity of applicants is desirable, name and demographic data may be deleted from resumes by someone not on the selection committee prior to the scoring of applicants. If few applicants are involved, personal interviews may also be considered. This does, of course, eliminate anonymity of the applicants.

After the attribute ranking has been completed, a total weighted attribute score may be computed for each applicant. The weighted attributes to be added are the products of attribute weights and attribute ranks.

5. Score Analysis

At this step, various forms of analysis may be employed, depending on the preference of the committee. One approach is:

a. Exclude the highest and lowest scores for each attribute. Then compute an average by attribute, using the remaining scores. For example, if scores of 18, 28, 30, 32, and 38 are stated for an attribute by committee members, the scores 18 and 38 are excluded when averaging. The result is an average score of 30 ((28 + 30 + 32)/3).

b. Compute a composite score for each applicant. This is accomplished by totaling the weighted attribute average scores.

c. Rank applicants in descending order, based on their composite score.

6. Verification
The committee should now evaluate certain applicants, preferably the top three through five, to confirm that the scoring technique has provided the quality of candidates desired. If the attribute identification and weighting have been accomplished properly, the committee will generally achieve consensus on the top three through five candidates. If not, re-evaluation of the attributes and weighting should be considered.

Choice

The committee should, after review of the resumes, select the candidate(s) to be personally interviewed. In cases where travel expenses are a constraint, the committee should interview the highest-scoring individual first. If the committee agrees that this person is satisfactory and the candidate accepts the position offered, the process is concluded. Otherwise, the remaining candidates should be interviewed in descending order until a suitable candidate has been determined who accepts the position.

An illustration of the selection technique, involving six applicants for a manager of programming position, is given in Figure 7-5. It uses the criteria shown in Figure 7-4. The scoring grid in Figure 7-5 consists of a matrix correlating each of the six applicants by attribute score, weighted score (WS), total weighted score (TWS) by category, composite score, and ranking.

The ranking of applicants by composite score indicates that Applicant F is the number-one candidate (quantitatively speaking), with a score of 348. Applicant C, with a score of 328, has scored high enough to warrant consideration also. The remaining applicants' scores are much lower. The committee should interview Applicant C and Applicant F and make a selection.

SUMMARY

Technology and personnel are components required to support an information system. The selection of technology and personnel should be based on their ability to contribute to the objectives of the information system. This selection is a decision-making process involving the three phases of intelligence, design, and choice.

	Education							*Work Experience*											*Management Experience*								
Attributes	EDP Training	W. S.	Bachelors	W. S.	Masters	W. S.	T. W. S.	Financial Apps.	W. S.	COBOL	W. S.	Assembler	W. S.	Online	W. S.	General EDP	W. S.	T. W. S.	Supervisory Resp.	W. S.	Professional Trng.	W. S.	Communication	W. S.	T. W. S.	Composite Score	Ranking
Weight	8		10		3			8		8		5		6		8			8		4		8				
Applicant A	4	32	4	40	5	15	87	2	16	4	32	3	15	0	0	3	24	87	3	24	2	8	3	24	56	230	5
Applicant B	5	40	5	50	0	0	90	4	32	5	40	5	25	3	18	5	40	155	0	0	0	0	2	16	16	261	4
Applicant C	5	40	5	50	0	0	90	5	40	5	40	2	10	4	24	5	40	154	4	32	3	12	5	40	84	328	2
Applicant D	3	24	3	30	0	0	54	3	24	3	24	2	10	0	0	3	24	82	2	16	1	4	1	8	28	164	6
Applicant E	4	32	4	40	1	3	75	4	32	5	40	3	15	2	12	4	32	131	4	32	4	16	3	24	72	278	3
Applicant F	5	40	5	50	5	15	105	4	32	5	40	3	15	4	24	5	40	151	5	40	3	12	5	40	92	348	1

Figure 7-5. *Scoring Grid for Applicants for a Manager of Programming for a Bank*

Intelligence pertaining to technology is available from in-house expertise, vendor representatives, consultants, literature, and other organizations. This intelligence is used to design alternatives through RFQ's or RFP's. The specifications defined in RFQ's and RFP's are categorized as mandatory and desirable, and assigned "hard-dollar" and "soft-dollar" values, respectively. The selection criteria for technology to be included in an information system should be based upon both cost and effectiveness.

Intelligence about the personnel required in an information system is based primarily on the technology used and the type of information processed. The design of alternatives and the actual selection of personnel can be accomplished using a quasi-quantitative selection technique in which selection attributes are defined, weighted, and scored for each applicant. This selection technique provides an effective, efficient means for selection decisions and comprehensive documentation for affirmative action and/or EEOC audits.

EXERCISES

1. Define and discuss potential sources of intelligence about hardware and software.
2. What is the difference between an RFQ and an RFP? When should each be used?
3. What are the possible consequences of having mandatory specifications that are really not necessary (i.e., that should be categorized as desirable specifications)?
4. What are "soft dollars" and why are they used?
5. Discuss the limitations of job descriptions as personnel selection criteria.
6. Within the personnel selection technique discussed in this chapter, what is the importance of weighting attributes for selection purposes?
7. Determine and price at least three competitive computer systems, supplied by different vendors, capable of meeting or exceeding the specifications listed below. If available, use DataPro or Auerbach as references. Otherwise, under the supervision of your instructor, contact three or more local computer vendor representatives to obtain the necessary information.

a. Hardware
   (1) Processor with a minimum 4 million bytes of memory.
   (2) Disk storage of at least 400 million characters with access time of 40 milliseconds or less.
   (3) Ability to support sixteen or more terminals.
   (4) Line printer capable of printing a minimum of 600 lines per minute (LPM).
   (5) Card reader capable of reading a minimum of 300 cards per minute (CPM).
   (6) Two tape drives: 40 KB with 800 BPI.
b. Software
   (1) Operating system must support multiprogramming.
   (2) Compilers must include COBOL and FORTRAN.
   (3) Utility programs must include sort and merge routines.

SELECTED REFERENCES

Bassler, Richard A., and Harold C. Demoody. *Computer System Evaluation and Selection—An Annotated Bibliography and Keyword Index.* Arlington, Va.: College Readings, 1971.

Cuthert, Norman H., and Janis M. Paterson. "Job Evaluation: Some Recent Thinking and Its Place in Investigation," *Personnel Management,* Sept 1966, pp. 156–162.

Equal Employment Opportunity Commission. *Fifth Annual Report.* Washington, D.C., 1971.

Ferrari, Domenico. "Workload Characterization and Selection in Computer Performance Measurement," *Computer* Vol. V (July–Aug 1972), pp. 18–24.

Fife, Dennis W. *Alternatives in Evaluation of Computer Systems.* (MTR-413). Bedford, Mass.: Mitre Corp., Dec 1968.

Goff, Norris F. "The Case for Benchmarking," *Computers and Automation,* May 1973, pp. 23–25.

Hillegass, John R. "Standardized Benchmark Problems Measure Computer Performance," *Computers and Automation,* Jan 1966, pp. 16–19.

Howard, Phillip C., ed. "Measuring System Performance with Benchmarks," *EDP Performance Review* Vol. L, No. 9 (Sept 1973), pp. 1–7.

Joslin, Edward O., ed. *Analysis, Design, and Selection of Computing Systems.* Arlington, Va.: College Readings, 1971.

Joslin, Edward O. *Computer Selection.* Reading, Mass.: Addison-Wesley, 1968.

McFarland, F. Warren; Richard L. Nolan; and David P. Norton. *Information Systems Administration.* New York: Holt, Rinehart and Winston, 1973.

Miner, John B., and Mary G. Miner. *Personnel and Industrial Relations, A Managerial Approach.* London: Collier-Macmillian Publishers, 1973.

Norman, A. K. *Industrial and Organization Psychology.* Englewood Cliffs, N.J.: Prentice-Hall, 1971.

Owens, W. A., and D. O. Jewell. "Personnel Selection," *Annual Review of Psychology,* 1969, pp. 26–29.

Simon, Herbert. *The New Science of Management Decision.* New York: Harper & Brothers, 1960.

Timmreck, Eric M. "Computer Selection Methodology," *Computing Surveys* Vol. 5, No. 4 (Dec 1973), pp. 199–222.

Timmreck, Eric M. "Performance Measurement: Vendor Specifications and Benchmarks," from *The Information Systems Handbook.* Homewood, Ill.: Dow Jones-Irwin, 1975, pp. 365–367.

Wetherbe, James C., and V. Thomas Dock. "Breaking the Description Dilemma: Personnel Selection by Group Analysis," *Data Management* Vol. 14, No. 12 (Dec 1976), pp. 16–19.

8 Systems Development, Implementation, and Evaluation

In this chapter, the final stages of the systems development cycle are discussed. Specific topics include the development, testing, implementation, and evaluation of information systems. These four processes pertain to the structuring of the technology and personnel included in an information system.

Development and testing are processes in which both programmers and systems analysts are actively involved. Implementation and evaluation are generally spearheaded by management and systems analysts. This chapter relates to the preceding three chapters as shown below:

	Inclusion	*Structure*
INFORMATION	Chapter 5	Chapter 6
TECHNOLOGY	Chapter 7	Chapter 8
PERSONNEL	Chapter 7	Chapter 8

DEVELOPMENT

Systems development pertains to structuring hardware and software to achieve effective and efficient processing of an information system. To a great extent, the structure of the hardware included in the system is determined as a result of the selection process discussed in Chapter 7. Hardware installation, which is usually performed by the vendor, may be all that is required. Therefore, for the systems analyst, development primarily pertains to the structuring or programming of application software to support the information system. Accordingly, the following discussion is directed at concepts and techniques of software development.

Problems in Software Development

Application programming has a rather controversial history. Such factors as shortages of computer programmers, frequent and enormous cost overruns on program development, and calamities caused by "bugs" residing in large, complex computer programs have done little to enhance the potential contributions of computer technology.

In most organizations, computer programming has not been subjected to traditional management measurements of quality and quantity. Rather, the mystique and glamour of computer technology has often caused management to take a "hands-off" approach to the specialized and often undisciplined activity of computer programming. The standards and quality control so essential to uniformity and reliability are often not enforced. Programs developed in such environments are difficult to understand, test, debug, and modify.

The computing industry is plagued with unfortunate stories of information systems that have been crippled or become useless with the turnover of critical programmers. In many of these cases, the program code was so complex and confusing that it was easier to write new programs than to figure out how the old ones worked.

Fortunately, considerable progress has been made in the area of computer programming standards and control. The major emerging concepts are:

- Structured programming[1]
- Top-down development[2]

1. Harlan D. Mills, "Top-Down Programming in Large Systems," *Debugging Techniques in Large Systems*, 1971, pp. 41–56.
2. Clement L. McGowan and John R. Kelly, *Top-Down Structured Programming Techniques*, 1975.

- Development support libraries[3]
- Chief programmer teams[4]

These concepts are discussed in the following subsections.

Structured Programming

Surprisingly, the programming of complex computations does not represent the major cause of software problems. Rather, the major contributor to confusing and error-prone program code is the nonstandard and undisciplined use of control (branching) logic. In such cases, the logic flow can erratically jump about in the program code, such that it is extremely difficult to follow.

The main culprit of undisciplined coding is the unrestricted use of unconditional branches (e.g., GO TO statements or their equivalents). The application of structured programming standards to control logic can resolve this problem. Programs can be written in such a fashion that they are straightforward and readable. The complexity and confusion involved in tracing undisciplined branching can be eliminated.

Structured programming is based on a solid theoretical foundation. Bohm and Jacopini have formally proved that only three basic control structures are necessary for computer programming.[5] These structures are:

1. The SEQUENTIAL structure—Statements are executed one after the other.
2. The DO WHILE structure—One or more statements are repeated as long as a condition is true.
3. The IF THEN ELSE structure—One of two statements is branched to, based on whether a given condition is true or false.

The logic within these three structures is presented in Figure 8-1. The three structures can be combined with each other to define program logic from beginning to end. Though unconditional branches are not necessary, on occasion they may be used for

---

3. Harlan D. Mills and John J. Naughton, "Programming Standards and Control," from *The Information Systems Handbook,* 1975, pp. 568–591.

4. IBM, "Improved Programming Technologies—An Overview," *Installation Management,* 1974, p. 3.

5. C. Bohm and G. Jacopini, "Flow Diagrams, Turing Machines, and Languages with Only Two Formation Rules," *Communications of the Association for Computing Machinery,* 1966, pp. 366–371.

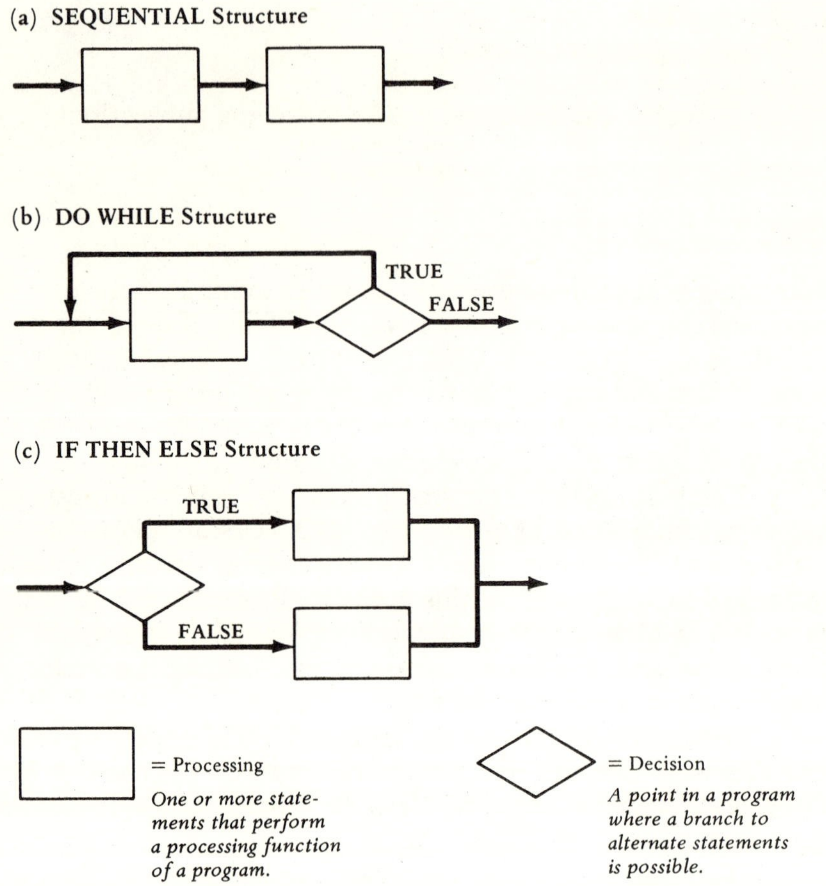

Figure 8-1. *Basic Control Structures of Structured Programming*

purposes of programming practicality. However, such cases should be made on an exception basis and carefully documented and justified.

Note that each of the structures has a single entry and a single exit. This single entry/exit property can be maintained as the structures are combined to build program modules or segments. For example, Figure 8-2 shows a logic flow that contains all of the structures and has the single entry/exit property.

The use of the single entry/exit structure in organizing logical structures allows the clustering of program functions into single entry/exit modules or subsystems. Assigning program functions to specific modules allows the modules of a complex program to be independently programmed by separate programmers. Because each module has only one point of entry and one point of exit, the interfacing of these modules into one program or system is simplified. Since each module's function is clearly de-

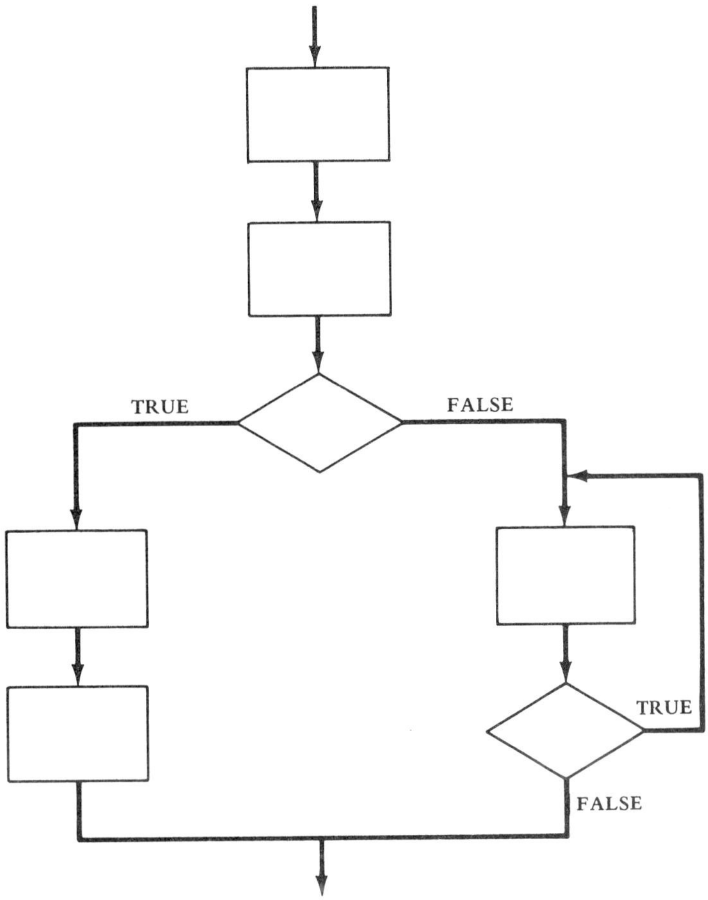

Figure 8-2. *Logic Flow with Single Entry/Exit and Multiple Control Structures*

fined before it is programmed, the modules fit together to form a complete program.

Besides allowing program functions to be programmed independently, modularization simplifies program testing, debugging, and modification. Since program functions are defined in specific modules, they can usually be tested, debugged, and modified as separate modules of program code. This reduces confusion caused by program statements branching into or out of modules at points other than where modules are designed to be entered or exited.

Top-Down Development

Program development has traditionally been a "bottom-up" procedure. The lowest-level functions are coded first. They are then tested and prepared for integration with other program functions[6]. This approach often results in data definition and interface problems during integration of the functions. Consequently, integration is delayed until these problems are resolved. Another major drawback of the bottom-up approach pertains to debugging program malfunctions after the functions have been integrated. It may be extremely difficult to determine which of the many program modules combined during integration is the source of a particular problem.

A bottom-up approach philosophically contradicts the manner in which a system evolves. The systems development process evolves from a broad and conceptual level to an increasingly detailed and operational level. Much of the low-level detail required for a bottom-up approach is not available until late in the systems development cycle.

"Top-down" program development is a procedure in which program requirements are factored from the top to the bottom. Top-down development is sometimes referred to as an outside/inside approach. The program to be developed is first considered in its broadest sense. This involves considering the program in terms of its inputs and outputs (i.e., what is to happen *outside* the program). The program is then factored into the progressively smaller functional units required to achieve the desired transformation of inputs to outputs (i.e., what is to happen *inside* the program).

Structured programming is a means for conveniently factoring large complex programs into manageable single entry/exit modules. The factoring continues until modules are defined at a level directly transferable to program code. The factoring process is illustrated in Figure 8-3. Each module (A through O in Figure 8-3) can be independently programmed, tested, debugged, and, if necessary, modified.

During the testing and debugging of a program module, other modules to be integrated may not yet exist. Dummy units called program stubs are used to simulate such modules. Program stubs do not usually perform meaningful computations. They may send acceptable responses back to another module so that it can continue processing as if the program stubs in fact executed as real modules. The eventual integration of the various program modules is not a problem because well-defined data definitions and interfaces are inherent to the top-down approach.

Figure 8-3. *Top-Down Approach to Developing a Computer Program*

---

6. IBM, "Improved Programming Technologies—An Overview," *Installation Management*, 1974, p. 3.

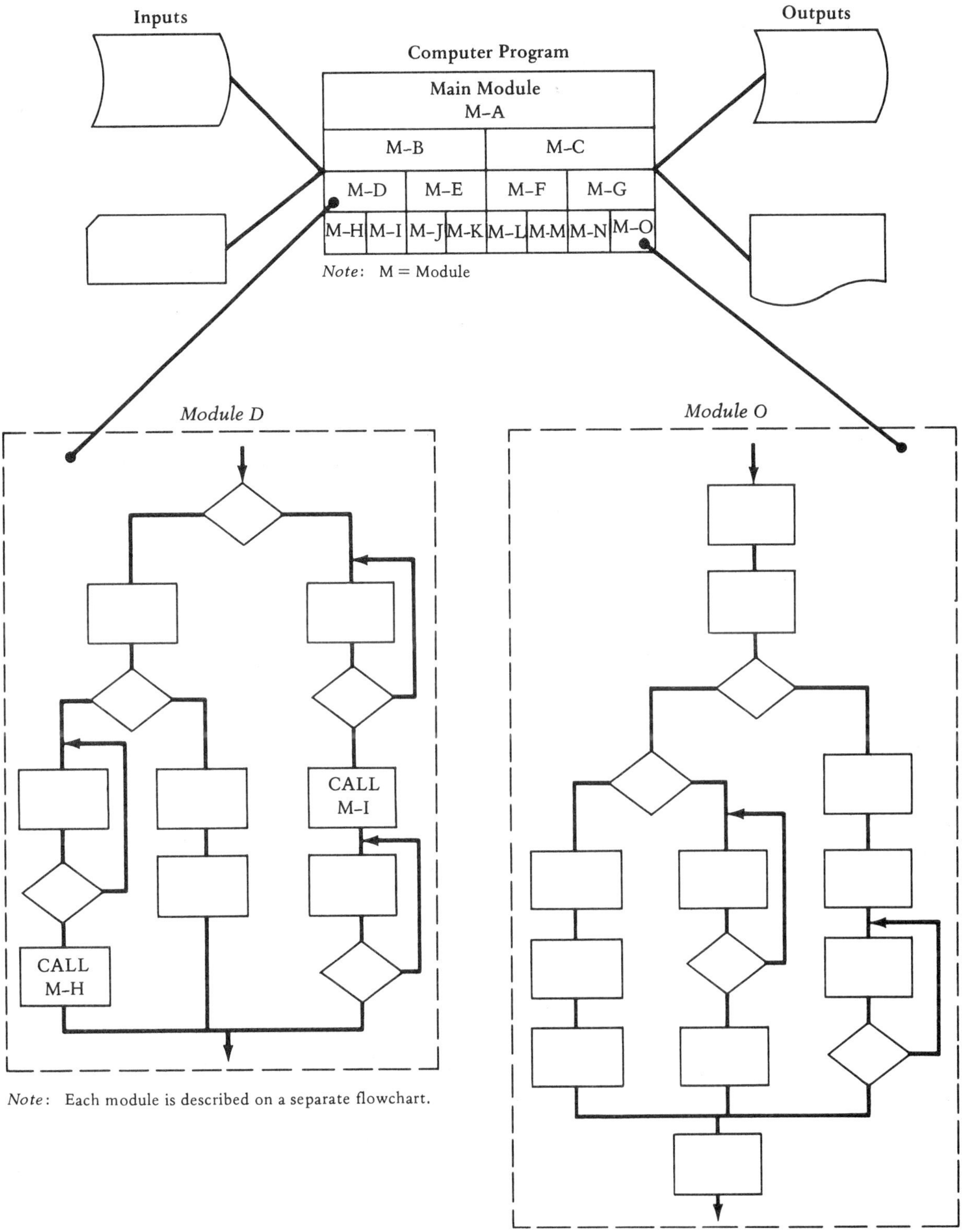

Note: Each module is described on a separate flowchart.

Three guidelines have proven useful for top-down program development. They are:

1. Use indentation in source code to highlight the DO WHILE and IF THEN ELSE structures.
2. Keep program modules to fifty or fewer statements.
3. A program module should be coded only after the higher-level module that invokes it has been coded and tested. (For example, in Figure 8-3, Module A should be coded and tested before Modules C and D.)

IBM has developed a system of procedures and forms called **HIPO** (**H**ierarchy plus **I**nput **P**rocess **O**utput) that is designed to facilitate effective and efficient top-down program development. HIPO provides excellent discipline and structure to factor complex programs into individual single entry/exit modules. Each module is defined in terms of its inputs, processes, and outputs.

HIPO provides a visual representation of software that serves as both a design tool and documentation. Completed HIPO forms (or charts) provide a blueprint for program code and reduce the need for program flowcharts. The HIPO charts partition all program modules, define the hierarchy and organization of the modules, and define the communication between modules. HIPO has been criticized because it does not usually define the sequence of execution (i.e., conditional branching or iterations) as flowcharts do. However, when necessary, HIPO specifications can be augmented with flowcharts.

Development Support Libraries

Traditionally, programmer productivity has been lowered because of the necessity to do clerical activities peripheral to program development. Such activities include filing and maintaining program listings and creating, updating, and backing-up program libraries. If such activities are not interfering with program development, they are often being neglected, which is even worse. Failure to perform these tasks jeopardizes the integrity and operation of an information system.

To address the issues of programmer productivity and integrity, many organizations have implemented a Development Support Library (DSL) function. A DSL is software for maintaining continuous documentation of major activities associated with developing a system. Operation of the DSL is usually the responsibility of one or more program librarians who are not under the supervision of those programming the system.

The DSL consists of two basic components: an internal library and an external library. The internal library is stored on computer files and contains all programs, test data, linkage-editing statements, and job control statements. The external library is maintained in human-readable form and describes the contents and the status of the internal library. Machine procedures are required to operate the internal library, and clerical procedures are required to operate the external library. The basic procedures required for each library are listed below:

Internal Library	*External Library*
1. Creating, updating, and deleting libraries	1. Documenting directions marked by programmer
2. Program module retrieval	2. Using internal library
3. Execution of programs	3. Filing status listings
4. Backing-up and restoring libraries	4. Filing and replacing pages in systems documentation
5. Producing status listings of library	

The DSL reduces the number of clerical activities that have to be done by programmers. It also defines sole responsiblity for the DSL functions. Accordingly, both programmer productivity and control of the development process are improved.

Chief Programmer Teams

Because of the increasing size and complexity of major programming efforts, a team effort is often required to complete a system on schedule. The coordinating and interfacing of the efforts of several computer programmers has traditionally proved to be a cumbersome task. The advent of structured programming, top-down design, and DSL's has provided the necessary framework for a team approach to program development. Coordination and integration of activities are simplified by the modularity of tasks that these interrelated techniques afford.

The use of a team of programmers headed by a chief programmer has proved to be highly productive. It is also a means of achieving high-quality program development.[7] The chief pro-

7. F. T. Baker, "Chief Programmer Team, Management of Production Programming," *IBM Systems Journal*, 1972, pp. 56–73.

grammer team concept involves a team effort. The team members perform specific functions, as follows:

1. Chief programmer—The chief programmer is a highly proficient programmer who has overall responsibility for the program design and coding. He or she defines and assigns program modules to be coded by other team members. He or she reviews all code and supervises the testing and debugging of code. The chief programmer informs management of project status and arranges for additional team members when necessary.
2. Backup programmer—The backup programmer is an experienced programmer who works closely enough with the chief programmer to assume his or her responsibilities, if necessary. He or she also participates in coding of program modules.
3. Program librarian—The program librarian is a clerical person who handles the DSL functions described in the preceding section, "Development Support Libraries."
4. Other—Additional team members (primarily programmers) are assigned to the team as necessary. These team members generally have specialized skills (e.g., coding speed, application knowledge, or knowledge of unique coding techniques).

Chief programmer teams provide increased control and integrity to the development process. They also provide for professional growth and technical excellence in programming. Clerical activities are done by the program librarian, thereby releasing programmers to devote their technical skills to developing programs. Inexperienced programmers are exposed to senior-level programmers in an environment that prepares them for leadership roles on future teams.

TESTING

During program development, computer programs and program modules are tested. As the programs are integrated into the total information system, overall testing of the system is possible.

Testing is a critical stage of the systems development cycle. It is the major checkpoint prior to the actual implementation of the system. The consequences of system malfunctions are minimized when they are discovered and resolved during testing rather than during implementation.

Validation

System problems are discovered by subjecting the system to extensive validation. The major system processes to be validated are:

1. Clerical processing—The data collection and preparation procedures must be performed correctly.
2. Input processing—Transactions must be properly checked for errors and must be applied to the right records in the right files.
3. Computational processing—The proper variables must be computed using the proper arithmetic.
4. Logic processing—Decision rules must be executed using the correct sequencing and branching.
5. File accesses—Records must be stored in and retrieved from the right locations.
6. Output processing—The correct variables must be printed in the right places on printouts and displays.

Validation of these processes can be accomplished through the processing of real or fabricated transactions that represent normal and abnormal conditions. The outputs resulting from transaction processing can be validated by checking to see if they are correct. For example, the amount due on a loan can be computed on a calculator and compared to a corresponding output provided by the computer system.

An information system must be able to handle exceptions (i.e., abnormal transactions). To validate this aspect of the system, transactions that deviate from normal transaction form should be prepared and processed. For example, payroll transactions with invalid department codes, missing social security numbers, pay rates below the minimum wage rate, or hours-worked that exceed the allowable limit should be prepared. A properly developed information system should detect such errors and report them on error reports.

Debugging

The process of correcting system malfunctions is called *debugging.* The source of an incorrect output can exist anywhere in an information system. The problem can occur during data collection or preparation, during input processing, in the computational or logical processes of one or more computer programs, or at any other point. Therefore, to debug a malfunction, the source of its occurrence must be isolated before the problem can be resolved.

Malfunctions occur somewhere between the initial entry of data to the system and the final output. In other words, if the data entering the system are correct, but the information leaving it is incorrect, then the malfunction resides somewhere between entry and exit. The specific location of the malfunction can be isolated by working backward from the output to the input and/or forward from the input to the output. Each point in the system where the data are used can be checked to determine where the malfunction occurred. That is, the data can be displayed before and after each system process until the specific process causing the problem is isolated. In many cases, data are not reported or displayed before and/or after a process (e.g., before and after a sort). In such cases, a special program may have to be written or a simple dump of the file taken for debugging purposes. Figure 8-4 illustrates a system with identified checkpoints for isolating a system malfunction.

Once isolated, a system malfunction can be resolved—say, by a change in program code, data collection procedure, or keypunch procedure. If the malfunction exists in a computer program, the use of structured programming facilitates debugging. Since the function of each module is defined, each module that uses the data in question can be determined and checked for malfunction. Once isolated, the program module(s) can be corrected and integrated back into the system.

IMPLEMENTATION

After development and testing has been completed, implementation of the information system can begin. During systems implementation, the project team (described in Chapter 5) should be brought back to full strength. After the problem and/or opportunity definition and analysis steps, project teams tend to play a rather passive role as the technical steps of systems design, component selection, program development, and testing evolve. However, the broad organizational representation, possible through the project team, is required to complete the systems development cycle.

It may be appropriate to reorganize and reorient the project team for completing what is now a much more refined and defined information system. More insight into the actual use and operation of the system is now possible and likely to affect the organizing and planning for implementation.

Another important reason for revitalizing the project team is that the interaction of the team members with their respective

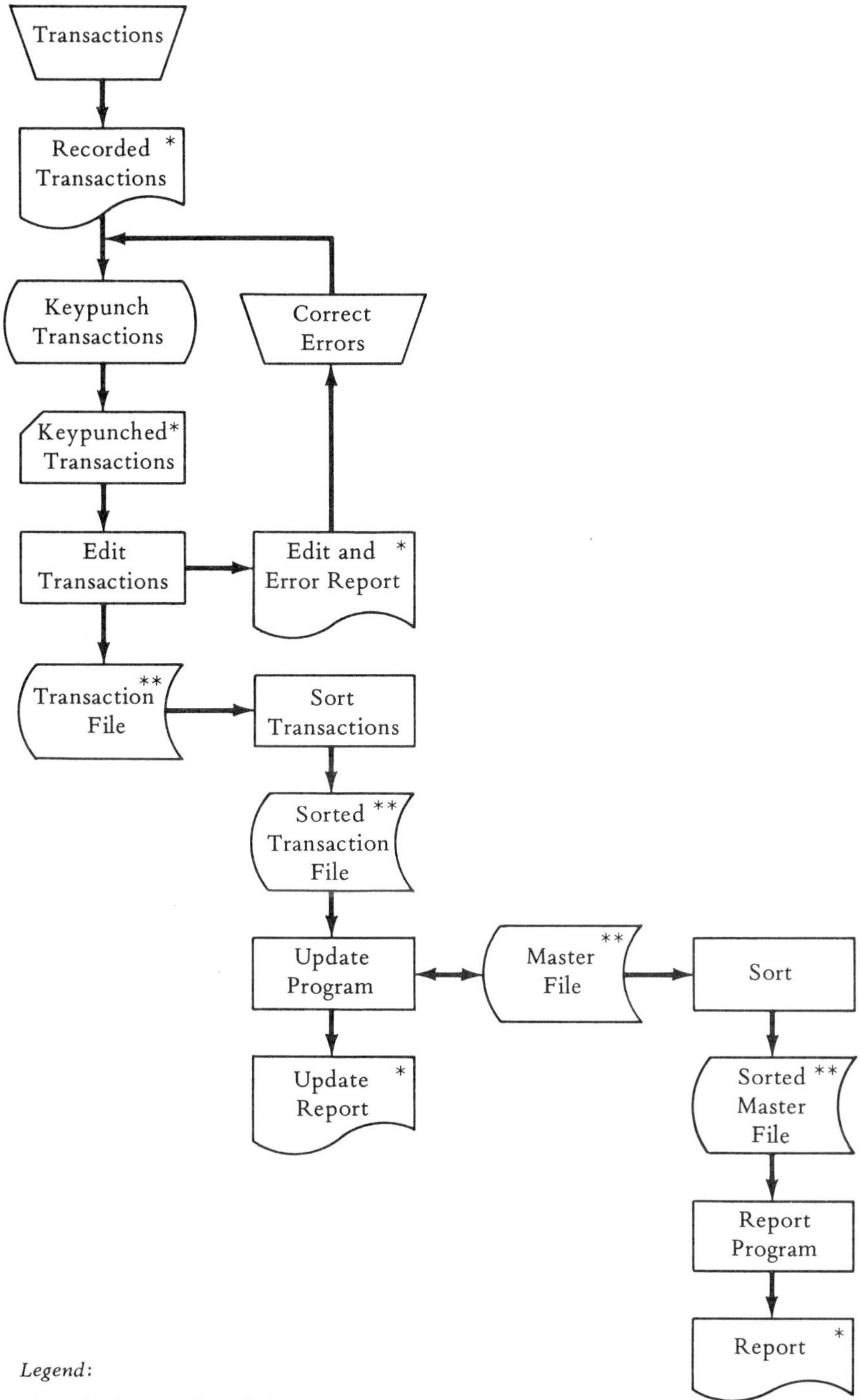

Legend:

* = Checkpoints for isolating system malfunctions.

** = Checkpoints requiring special programs or file dumps.

Figure 8-4. *Checkpoints for Debugging an Information System*

departments is important to facilitate communications concerning the forthcoming changes. Members of user departments tend to be more comfortable and cooperative if they participate, or feel they are adequately represented, during the implementation of a new information system.

User Training

Thorough user training is critical to the successful implementation of an information system. User training requirements for a new information system can be categorized as either clerical or managerial:

1. Clerical—Must be instructed in the processing of transactions. This may involve training in the operation of computer terminals and/or the filling out of documents that can be converted to machine-readable form.
2. Managerial—Must be informed as to the formats and contents of reports and terminal displays and how to request reports or make terminal inquiries. Managers also need to understand how programmed decisions are made and how to use any decision-support capabilities.

It is often advantageous to have representatives (possibly project team members) from the different departments conduct training sessions within their respective departments. Departmental representatives should also participate in the development of user documentation (e.g., user procedures manuals) to be used in operating the system. Clerical and managerial personnel are often less intimidated by computer technology if a new information system is explained to them and documented for them by someone from their own department rather than by a computer technician.

If an information system affects several departments and a large number of employees, it may be advisable to formalize user training requirements in a user training checklist to insure complete coverage. An example of a user training checklist is given in Figure 8-5. Note that the specifics of training, the responsibility, and the time frame are clearly defined.

Frequently, user training overlaps program development. Much of the required training can be accomplished without the use of an operational system. Also, systems tend to be implemented in a modular fashion (e.g., the inventory module of an order processing system may be operational before the order entry

Figure 8-5. *User Training Checklist*

Training Checklist
for New Order Processing System

Inventory Subsystem

Description of Activity	*Individual Responsible*	*Department Responsible*	*Start Date*	*Completion Date*	*Approved By*
Management Training					
Overview	J. Johnson	Plant Mgr.	9/1	9/1	
Information Available	B. Benton	Inventory	9/5	9/7	
Programmed Decisions	B. Benton	Inventory	9/5	9/7	
Decision-Support Capabilities	H. Hendrix	Info. Systems	9/8	9/8	
Feedback Procedures	R. Scammel	Info. Systems	9/8	9/8	
Clerical Training					
Overview	J. Johnson	Plant Mgr.	9/2	9/2	
Document Preparation & Handling	C. Carnes	Inventory	9/5	9/10	
Terminal Operation	C. Carnes	Inventory	9/10	9/30	

Accounts Receivable Subsystem

Description of Activity	*Individual Responsible*	*Department Responsible*	*Start Date*	*Completion Date*	*Approved By*
Management Training					
Overview	B. Barnes	Finance	9/20	9/20	
Information Available	B. Black	Accounting	9/25	9/26	
Programmed Decisions	S. Smith	Credit	9/27	9/27	
Decision-Support Capabilities	B. Black	Accounting	9/28	9/28	
Feedback Procedures	R. Scammel	Info. Systems	9/28	9/28	
Clerical Training					
Overview	B. Barnes	Finance	9/21	9/21	
Document Preparation & Handling	S. Sharp	Credit	9/25	9/27	
Terminal Operation	S. Sharp	Credit	10/1	10/15	

Order Entry Subsystem

Description of Activity	*Individual Responsible*	*Department Responsible*	*Start Date*	*Completion Date*	*Approved By*
Management Training					
Overview	G. Gordon	Sales	10/1	10/1	
Information Available	J. Jackson	Order Processing	10/2	10/5	
Programmed Decisions	J. Jackson	Order Processing	10/6	10/6	
Decision-Support Capabilities	G. Gordon	Sales	10/10	10/15	
Feedback Procedures	R. Scammel	Info. Systems	10/15	10/15	
Clerical Training					
Overview	J. Jackson	Order Processing	10/2	10/2	
Document Preparation & Handling	W. Wilson	Order Processing	10/5	10/7	
Terminal Operation	W. Wilson	Order Processing	10/8	10/20	

module). This allows operational training to be staggered in the sequence in which modules are operationalized. To the extent that an early start on user training is possible, it is generally wise to proceed.

Conversion

Conversion is the process of changing from the old system to the new system. Conversion requires careful planning to establish the basic approach to be used well in advance of the actual cutover. Two major planning decisions that establish the basic conversion methodology pertain to the degree of parallel processing to be used and the sequence of conversion activities.

Parallel and Discrete Processing

Conversion methodology can be defined in terms of a continuum ranging from parallel to discrete. In a parallel conversion, both the old and the new systems are concurrently processed until the new system stabilizes. This insures the reliability of the new system prior to abandoning the old one. In a discrete conversion, the cutover to a new system terminates the use of the old system.

The conversion approach may reside somewhere between parallel and discrete. For example, transactions processed on the old system can be collected for a week and then reprocessed on the new system over the weekend. In this example, parallel processing applies to the transactions processed but not to the time frame in which they are processed.

Parallel processing reduces the risk of implementing a new system. However, it has disadvantages. First, parallel processing is expensive. Often, additional personnel (or overtime) and/or equipment are required. For example, assume a wholesale distributor is converting from batch to online processing of customer orders. Parallel processing during the implementation requires preparing the transaction documents for keypunching (the old system) in addition to entering the transactions through terminals (the new system). Card-punch machines and card-punch personnel will still be required as will the computer-processing and report-generating functions of the old system. If the computer that supports the online processing is to replace an old computer, both the new and the old computers are required until conversion is completed and parallel processing is terminated.

For some systems, it is virtually impossible to conduct parallel processing. For example, it is not plausible to parallel process two online airline reservation systems. In discrete conversion, rigorous testing of the new system is critical prior to implementation. Where parallel processing is possible, value judgments must be made to assess the tradeoffs between risk and costs.

Conversion Sequence

In implementing a new information system, it is important to determine the sequence in which system segments are to be converted. The ordering of the conversion sequence is generally a function of the following variables:

- Organizational functions—The different departments, branches, plants, or other logical breakdowns.
- Organizational cycles—The normal cycle of organizational activities within organizational functions (e.g., accounting or billing cycles).

It is generally advisable to begin conversion with segments of the system that appear to offer the least difficulty. For example, the conversion may start with the best-managed and best-organized functions with the shortest cycles. This enhances the probability of initial successes, which in turn enhances the credibility and momentum of the new system and its implementation.

During the conversion, management should attempt to minimize any additional internal or external requirements on the department undergoing conversion. A department is generally disrupted enough during conversion without having to handle the complications of additional disturbances.

A checklist of conversion activities and their sequence should be made to insure that all bases are covered. Such a document defines responsibilities and schedule dates for conversion activities in much the same manner as the user training checklist defines training requirements (look again at Figure 8-5).

The final task of the conversion sequence is to define what will signify the complete implementation of the new system. This definition may vary. Whatever the definition of implementation, it should be agreed upon in advance by organizational members. When the defined implementation has been achieved, the system should be formally approved and accepted, and implementation should be considered completed.

EVALUATION

The final step (and, one of the most important steps) of the systems development cycle is evaluation. Evaluation provides the feedback necessary to assess the value of the information and the performance of the personnel and technology included in the information system. This feedback serves two functions. First, it provides information as to what adjustments to the information system may be necessary. Second, it provides information as to what adjustments should be made in approaching future information systems development projects.

There are two basic dimensions of information systems that should be evaluated. One dimension is concerned with whether or not the system was developed and operates properly. The other dimension is concerned with whether or not the proper information system was developed.

Development Evaluation

Evaluation of the development process is primarily concerned with whether the system was developed on schedule and within budget. This is a rather straightforward evaluation. However, it requires that schedules and budgets be established in advance and that records be kept of actual performance and costs. Procedures for scheduling and budgeting information systems development are discussed in the next chapter, "Systems Administration."

Historically, few information systems have been developed on schedule and within budget. In fact, many information systems have been developed without clearly defined schedules or budgets. The mystique and uncertainty associated with information systems development has often resulted in its not being subjected to traditional management control procedures. Fortunately, the increasing experience base and better understanding of systems development by both technicians and management have resulted in a greater emphasis on planning and control of information systems development.

Operation Evaluation

The evaluation of the operation of the information system pertains to whether the hardware, software, and personnel are capable and do, in fact, perform as they are supposed to. Operation evaluation answers such questions as:

1. Are all transactions processed on time?
2. Are all values computed accurately?
3. Is the system easy to work with and understand?
4. Is terminal response time within acceptable limits?
5. Are reports processed on time?
6. Is there adequate storage capacity for data?

Operational evaluation (like development evaluation) tends to be relatively straightforward if evaluation criteria are established in advance. For example, assume a system is supposed to be capable of supporting 100 terminals with less than a 2-second response time. Evaluation of this aspect of system operation is easy after the system is operational.

Information Evaluation

An information system may perform exceptionally. Computations may be correct; transactions may be efficiently processed; and reports may be distributed on time. However, a properly functioning system may not be a system that is proper for the needs of an organization. An information system should also be evaluated in terms of the information it provides. This aspect of information system evaluation is difficult. It cannot be conducted in the straightforward quantitative manner that is possible for development and operation evaluations.

The objective of an information system is to provide information to support the organizational decision system. Therefore, the extent to which information is relative or not relative to decision making is the area of concern in evaluating the information provided by an information system. However, it is a practical impossibility to evaluate directly an information system's support of decisions made in an organization.

A viable approach for indirectly measuring and evaluating the information provided by an information system has been proposed by Richard L. Nolan and Henry H. Seward.[8] Their approach is based on the concept that the more frequently a decision maker's information needs are met by an information system, the more satisfied he or she tends to be with the information system. Conversely, the more frequently necessary information is not available, the greater the effort (and frustration) required to obtain the necessary information, and hence the greater the dissatisfac-

8. Richard L. Nolan and Henry H. Seward, "Measuring User Satisfaction to Evaluate Information Systems," *Managing the Data Resource Function*, 1974, pp. 253–276.

tion with the information system. This concept is illustrated in Figure 8-6. Since satisfaction with an information system correlates with the ability of the information system to support decision making, satisfaction can be used as a surrogate to evaluate the information provided by an information system.

Measurement of user satisfaction can be accomplished using the interview and questionnaire techniques discussed in Chapter 5. If management is generally satisfied with an information system, it is reasonable to assume that the system is meeting the organization's requirements. If management is not satisfied, modifications ranging from minor adjustments to complete redesign may be required.

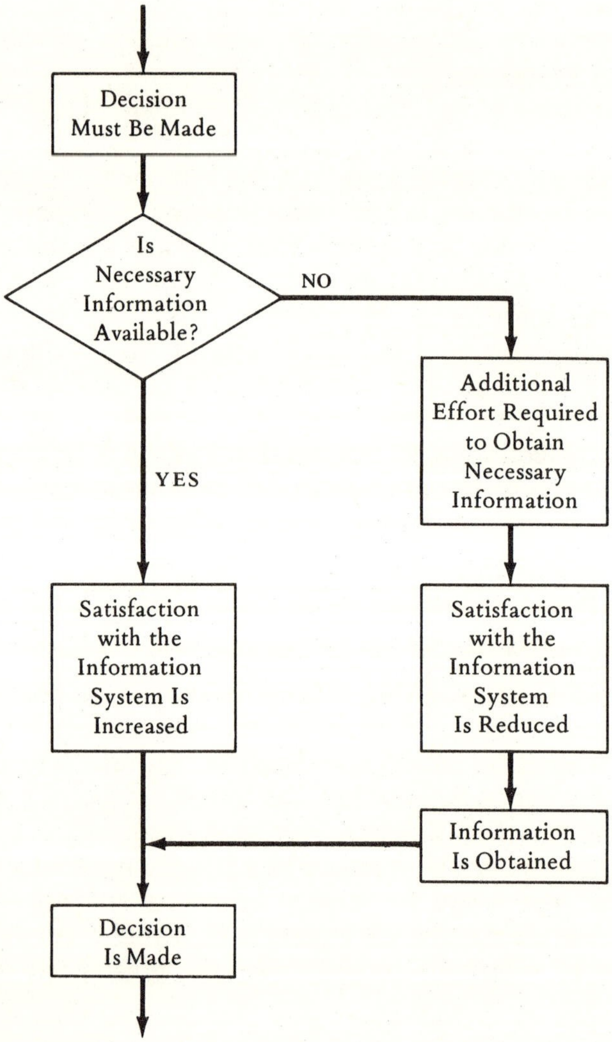

Figure 8-6. *Relationship between Information System Support of Decision Making and User Satisfaction*

Maintenance

Most information systems require at least some modifications after development. This need arises from a failure to anticipate all requirements during systems design and/or from changing organizational requirements. The second of these, changing requirements, continues to impact most information systems as long as they are in operation. Consequently, periodic systems maintenance is a requirement of most information systems. Systems maintenance involves adding new data elements, modifying reports, adding new reports, changing calculations (e.g., payroll tax tables), and the like.

Maintenance can be categorized as either scheduled maintenance or rescue maintenance. Scheduled maintenance is anticipated and can be planned for. For example, the implementation of a new inventory coding scheme can be planned in advance. Rescue maintenance pertains to previously undetected malfunctions that are not anticipated but that require immediate resolution. A system that is properly developed and tested should have few occasions of rescue maintenance.

An information system may remain in an operational and maintenance mode for several years. The system should be evaluated periodically to insure that it is operating properly and is still proper for the organization. When a system becomes a problem (i.e., no longer satisfies the organization's needs) or new opportunities are available (e.g., new technology), an information system may be replaced by a new information system, generated from a new information systems development cycle.

SUMMARY

The final stages of the systems development cycle include systems development, testing, implementation, and evaluation. They are primarily involved with the structuring of technology and personnel included in the information system.

Systems development is concerned with the computer programming necessary to process the data in the information system. Four useful techniques for managing and controlling program development are structured programming, top-down development, Development Support Libraries, and chief programmer teams. Once developed, a system should be tested by validation techniques and by isolating and debugging any malfunctions.

After thorough testing, an information system can be implemented. Implementation involves management and clerical training in the use of the system, followed by conversion from the old system to the new one. The conversion process may or may not employ parallel processing. In any case, the conversion sequence should be clearly defined.

After the system has been implemented, it should be evaluated. Evaluation is concerned with determining whether the system works right (i.e., was developed and operates properly) and whether it is the right system (i.e., provides the proper information). Once evaluated, an information system remains in a maintenance mode until it is replaced by a new information system.

EXERCISES

1. Discuss the complications caused by unconditional branches in a computer program.
2. Flowchart the following sequences and define which of the three basic control structures from Figure 8-1 apply:
   a. (1) A = B + C
      (2) D = A + 5
      (3) E = D/A
   b. (1) A = A + 1
      (2) Repeat the preceding step until A = 10, then add A to B, giving C.
   c. (1) A = B + C
      (2) If A is negative, multiply it by −1; otherwise, divide it by 5.
3. Using structured programming control logic, flowchart the following program module with a single entry/exit and no unconditional branching.

   Problem Description

   This program module is to read transactions, each of which contains a customer account number (C-ACCT) and a charge (CHARGE) made by that customer. More than one transaction may be entered for a customer. The transactions are in account number order.

   As the program module reads transactions, it is to add charges to individual customer totals (C-TOT) and to a grand total (G-TOT). Each individual customer's total is to be printed after it is tallied. When the end of the transaction file is reached, the grand total is to be printed. The program module has then completed processing. HINTS: All totals must be initialized to 0. Conditional branches can branch to other conditional branches, and two or more branches can branch to the same place.

4. Discuss the disadvantages of bottom-up program development. How are these disadvantages overcome by the top-down approach?

5. Discuss the work of the backup programmer in a chief programmer team.

6. Why should testing of a system include abnormal as well as normal transactions?

7. The transaction validation rules for a computer program in an accounts receivable system are as follow:

   Account Number—Located in card columns 1–6. Must be numeric.
   Customer Name—Located in card columns 7–27. Can contain alphabetic characters or blanks.
   Store Number—Located in card columns 28–30. Must be a number between 99 and 201.
   Amount Charged—Located in card columns 31–37. Must be all numeric except for a decimal point in column 35. Cannot exceed 1000 or be negative.

   Create at least ten transactions that can be used to validate the input-error-detection module(s).

8. Why is structured programming helpful for both preventing program malfunctions and isolating them if they occur?

9. Discuss how to debug a system containing several programs, reports, and sorts.

10. Discuss the importance of reactivating the project team prior to implementation.

11. When is parallel processing not practical? Give examples.

12. Discuss how management can evaluate the value of the information provided by an information system.

13. Discuss how a system can work right but not be the right system, and vice versa.

SELECTED REFERENCES

Baker, F. T. "Chief Programmer Team, Management of Production Programming," *IBM Systems Journal* Vol. 11, No. 1 (Mar 1972), pp. 56–73.

Baker, F. T. "System Quality through Structured Programming," Fall Joint Computer Conference, 1972.

Bohm, C., and G. Jacopini. "Flow Diagrams, Turing Machines, and Languages with Only Two Formation Rules," *Communications of the ACM* Vol. 9, No. 5 (May 1966), pp. 366–371.

Dahl, O. J.; E. W. Dijkstra; and C. A. Hoare. *Structured Programming.* London: Academic Press, 1972.

Dijkstra, E. W. *Structured Programming.* New York: Academic Press, 1972.

Elspas, B.; K. N. Levitt; R. J. Waldinger; and A. Waksman. "An Assessment of Techniques for Proving Program Correctness," *ACM Computing Surveys* Vol. 4, No. 2 (June 1972), pp. 97–147.

IBM. "Improved Programming Technologies—An Overview," *Installation Management.* White Plains, N.Y.: IBM, 1974, p. 3.

Knutsen, K. Eric. "Business Systems Analysis: Program Design," from *The Information Systems Handbook.* ed. McFarlan and Nolan. Homewood, Ill.: Dow-Jones-Irwin, 1975, pp. 539–567.

Knutsen, K. Eric, and Richard L. Nolan. "On Cost/Benefit of Computer-Based Systems," *Managing the Data Resource Function.* ed. Nolan. St. Paul, Minn.: West, 1974, pp. 253–276.

McGowan, Clement L., and John R. Kelly. *Top-Down Structured Programming Techniques.* New York: Mason/Charter Publishers, 1975.

Mills, Harlan D. *Mathematical Foundations of Structured Programming.* IBM Report Number FSC 72-6012. 1972.

Mills, Harlan D. "Top Down Programming in Large Systems," in *Debugging Techniques in Large Systems.* ed. Randall Rustin. Courant Computer Science Symposium 1, NYU: 1971, pp. 41–56.

Mills, Harlan D., and John J. Naughton. "Programming Standards and Control," from *The Information Systems Handbook*. ed. McFarlan and Nolan. Homewood, Ill.: Dow-Jones-Irwin, 1975, pp. 568–591.

Nolan, Richard L., and Henry H. Seward. "Measuring User Satisfaction to Evaluate Information Systems," in *Managing the Data Resource Function*. ed. Nolan. St. Paul, Minn.: West, 1974, pp. 253–276.

Parnas, D. L. "On the Criteria to be Used in Decomposing Systems into Modules," *Communications of the ACM* Vol. 15, No. 12 (Dec 1972), pp. 1053–1058.

Sammet, J. C. "Perspective on Methods of Improving Software Development," *Software Engineering* Vol. 1, 1970.

Weinberg, Gerald M. *The Psychology of Computer Programming.* New York: Van Nostrand Reinhold Co., 1971.

9 Systems Administration

The commitment to information systems projects and the resulting systems development cycles must be properly administered under management's direction. In this chapter, basic concepts and approaches to information systems administration are discussed.

Though systems analysts are not directly involved in the overall administration of information systems, they should be knowledgeable of systems administration processes for the following reasons:

1. Systems analysts are subject to the procedures and controls resulting from systems administration. Therefore, their effectiveness is increased if they understand systems administration.
2. Systems analysts can often make suggestions that constructively influence top management's approach to systems administration. This is particularly true if management is inexperienced in systems administration.
3. Systems analysts often aspire to move into management. An understanding and demonstration of administrative skills can facilitate such advancement.

This chapter is organized into four sections: historical perspective, organization, planning and control, and behavioral considerations. The relationship of this chapter to the preceding four chapters is illustrated at the top of the next page.

HISTORICAL PERSPECTIVE

The results of computer-based information systems in most organizations have fallen substantially below management's original expectations. Most information systems projects have failed to deliver what was promised for the cost and/or in the time-frame expected. For example, Weyerhauser Co., the far-flung forest products complex, has probably made as effective use of computer-based information systems as any other U.S. corporation.[1] How-

1. *Business Week* Editorial, "Business Takes a Second Look at Computers," *Business Week*, June 5, 1971, p. 63.

Chapter 9

	Inclusion	*Structure*
INFORMATION	Chapter 5	Chapter 6
TECHNOLOGY	Chapter 7	Chapter 8
PERSONNEL	Chapter 7	Chapter 8

ever, they have experienced their disappointments. When they began to develop a new online inventory system, they were led to believe they could do it in one year with ten people. Instead, the project took three years and fifty people. Such experiences are not uncommon.

A partial, but unfortunate indicator of the failure of many computing efforts is the high turnover of data processing and information systems managers—as high as 40 percent to 50 percent annually. Further contributing to management's disenchantment with the computing effort is the increasingly high cost associated with it. Prior to 1960, computer-based information systems were virtually nonexistent in most organizations. Today, large organizations frequently spend millions of dollars annually on this new technology.

DIFFICULTIES OF PLANNING AND CONTROL

Perhaps the most common management criterion for evaluating the success or failure of a project is the degree of variation between planned performance and actual performance. However, to be able to compare planned performance with actual performance, a plan that defines the objectives the project is to achieve and the criteria for measuring achievement must be established. It is unfair and even unrealistic of management to be critical of computing achievements unless management's expectations have been clearly articulated in planning documents that can be used to properly direct the computing effort. Ironically, the computing effort has been conspicuously void of planning and control activities in many organizations. Organizations that do plan and control information systems can expect to enjoy greater success with their computing efforts.[2]

In view of the overwhelming impact of computers on organizations, it is alarming to see them undergo less management scrutiny than other functions. However, closer analysis into the uniqueness of computing provides some insight into managements' reluctance and/or inability to cope with planning and controlling the computing effort. Computing as an organizational activity is complicated due to the following factors:[3,4]

1. The computing resource is integrated into virtually every dimension of modern organizations in both explicit and obscure ways. Indeed, it has become increasingly difficult to define exactly where computing activities begin and other activities leave off.
2. The computer resource has a complex set of supply/demand characteristics:
   a. The ratio of fixed to variable costs is high.
   b. Computer hardware usually offer economies of scale.
   c. Incremental computing capacity often must be acquired in large blocks.
   d. Needs for EDP services grow rapidly in complexity and sheer size.
   e. Processing tends to be cyclical.

---

2. McKinsey & Company, Inc., "Unlocking the Computer's Profit Potential," *The McKinsey Quarterly*, Fall 1968, pp. 17–31.

3. John Dearden and Richard Nolan, "How to Control the Computer Resource," *Harvard Business Review*, Nov—Dec 1973, pp. 68–78.

4. F. Warren McFarland, "Management Audit of the EDP Department," *Harvard Business Review*, May—June, 1973, pp. 131–142.

f. One computer system is frequently unable to serve all of the diverse demands that a large organization can place on it.
g. Processing priorities are highly variable, depending on the application, the users. and the timing.

3. Computing technology is extremely dynamic and is changing at an accelerating rate. The economical and technical feasibility of new computer applications is continually improving, resulting in a proliferation of additional computer applications.
4. Staffing creates considerable uncertainty. Personnel requirements are often as dynamic as the computing industry itself. Skills and experience that were highly valued ten years ago may be obsolete and even potentially damaging today.

Four Stages of EDP Growth

Considerable enlightenment into both the impact of computer-based information systems and the need for improved organizing, planning, and control has resulted from research efforts headed by Richard L. Nolan.[5] His and his colleagues' research findings indicate that organizations go through the four stages of EDP growth defined below:

Stage 1: *Initiation.* Development of cost-reduction accounting applications.
Stage 2: *Expansion.* Proliferation of applications in all functional areas of the organization.
Stage 3: *Formalization.* Moratorium on new applications with emphasis on planning and control of future information systems.
Stage 4: *Maturity.* Implementation of database or integrated applications.

Characteristics of organization and of planning and control during each of the four stages are illustrated in Table 9-1.

This four-stage framework can help management and information systems personnel to understand the current status of their organization and the likely course of future developments. The successive stages indicate a higher level of maturity and

5. Richard L. Nolan and Cyrus F. Gibson, "Managing the Four Stages of EDP Growth," *Harvard Business Review*, Jan–Feb 1974, pp. 76–88.

Table 9-1. *Characteristics of Organization and of Planning and Control during the Four Stages of EDP Growth*

	Stage 1 Initiation	*Stage 2* Expansion	*Stage 3* Formalization	*Stage 4* Maturity
ORGANIZATION	The data processing or information systems function is organized under the department of first-applications justification; it is generally a small department.	The data processing or information systems manager is moved up in the organization; systems analysts and programmers are assigned to work in the various functional areas.	The data processing or information systems function moves out of the functional area of first applications; a steering committee is set up; control is exerted through centralization; maintenance programming and systems programming become dominant activities.	The data processing or information systems function is set up as a separate area; the data processing or information systems manager takes on a higher-level position; some systems analysts and (sometimes) programmers are decentralized to user areas; high specialization appears in computer configuration and operation; systems design and programming take on a consulting role.

sophistication in the computing effort. They are usually increasingly difficult to achieve. The systems development cycle in each of the four stages remains essentially the same, though it is generally more refined and polished as an organization advances through the four stages.

The costs incurred during these four stages are usually expressed in the form of an S-shaped curve (see Figure 9-1). It is not until the third stage of EDP growth that management planning and control becomes a major area of concern. It is generally not until this stage that the combination of increasing costs and project failure generates enough concern to invoke management action. It is during and after the third stage that increasing costs come under control.

Table 9-1. *(cont'd)*

	Stage 1 Initiation	*Stage 2* Expansion	*Stage 3* Formalization	*Stage 4* Maturity
PLANNING AND CONTROL	Loose planning and budgeting.	Loose planning and budgeting.	Strong budgetary planning for hardware facilities and new applications.	Multiple 3–5 year plans for hardware, facilities, personnel, and new applications.
	Controls notably lacking; priorities assigned on a "first come, first serve" basis with no charge for services.	Lax controls, intended to engender applications development; few standards, informal project control.	Proliferation of controls to contain a runaway budget; formal priority setting; budget justification. Programming controls; documentation; standards. Project management initiated; management reporting system introduced: project plan, project performance, customer service, personnel resources, equipment resources, budget performance. Chargeout introduced; post-system audits. Quality control policies for computer system, systems design, programming, operations.	Refinement of management control—elimination of ineffective control techniques and further development of others; introduction of database policies and standards; focus on pricing of computer services for engendering effective use of the computer.

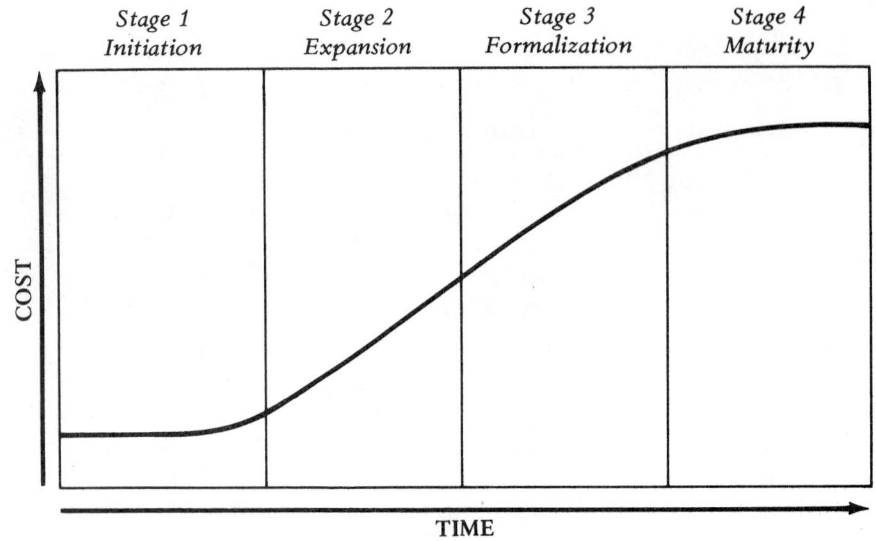

Figure 9-1. *Costs Incurred during the Four Stages of EDP Growth*

The concepts and approaches discussed in the next two sections of this chapter are aimed at providing the necessary framework to achieve the third and fourth stages of EDP growth. The fourth stage of EDP growth pertains to the development of integrated information systems using database technology. Material related to this stage is discussed in Chapter 10, "Management Information Systems and Database Management Systems."

ORGANIZATION

The three major organizational issues pertaining to the data processing or information systems function are:

1. Centralization versus decentralization of technology, personnel, and authority.
2. The location and reporting relationship of the data processing or information systems function.
3. The internal structure of the data processing or information systems function.

The third issue is discussed under "Organization and Management of Data Processing Personnel" in Chapter 4. The first two issues are discussed below.

Centralization versus Decentralization

Centralization or decentralization of organizational functions was a widely discussed and hotly debated topic long before computer technology entered the scene. However, the sudden, permeating effect that computer technology has had on organizational operations has highlighted it as an issue of discussion and debate.

Computer technology has introduced more than one dimension to the centralization/decentralization issue. Some aspects of the data processing or information systems function can be centralized and others decentralized. For example, computing equipment can be centralized at a large corporate computer center, and systems analysts and programmers can be decentralized into operating departments. Conversely, systems analysts and programmers can be centralized, and computing equipment can be decentralized.

There is no clear-cut position on the centralization/decentralization issue. Proponents of both sides can offer convincing arguments and successful examples of their respective views. It is, therefore, appropriate to discuss the commonly argued advantages of each.[6] (The disadvantages of one approach are implied by the advantages of the other.)

Arguments for Centralization

Arguments for centralization follow.

1. **Organization-wide consolidation of operating results.** Financial and operating data are readily consolidated for reporting and evaluation purposes. Without centralization, consolidation is usually obstructed by incompatibilities of different systems design, coding schemes, and data formats.
2. **Economies of scale.** Traditionally, the computer processing capabilities of large systems have increased geometrically as their costs increased arithmetically. Therefore, economies of scale have resulted when several small computers were replaced by one large computer. Economies of scale also apply to centralization of staff. Efficiencies are afforded by the reduction in duplication of effort and better control of the allocation of systems analysis and programming activities.
3. **Shortage of data processing or information systems personnel.** Computer professionals are scarce and often have

---

6. George Glaser, "The Centralization vs. Decentralization Issue: Arguments, Alternatives, and Guidelines," *Data Base,* 1970, pp. 1–7.

more allegiance to their profession than to the organizations for which they work. Consequently, turnover tends to be high, which is potentially damaging to an organization. Centralization reduces the impact of both shortages and turnovers by permitting one staff of larger size. That is, centralization of staff reduces the dependence on a few individuals. Recruiting and retainment of computer professionals are enhanced because personnel have greater opportunity to interact with a broader variety of staff. Consequently, they work in a more expansive environment.

4. **Ease of control.** Top management can more easily control operating divisions when the divisions use uniform information reporting systems. When divisions individually develop their own reporting systems, there are usually discrepancies in data used, data definitions, and reporting formats. Through centralized systems development, uniformity can be enforced.

Arguments for Decentralization

Arguments for decentralization follow.

1. **Familiarity with local problems.** The closer computer professionals are to problems, the more likely they are to devise good solutions. Centralization of equipment and staff tends to isolate them from the problems of the organizational functions they are attempting to support.
2. **Rapid response to local processing needs.** Computer equipment and staff can be more responsive to both development and production requirements when they are decentralized to user departments. User departments have more discretion in scheduling their resources. They do not have to compete with other departments for centralized computing services.
3. **Profit and loss responsibility.** When computer equipment and staff are decentralized to user departments, their costs can also be easily decentralized to users. This tends to make the user departments more sensitive to cost/benefit considerations because the computing costs directly affect their department profitability.

Guidelines

The preceding discussion indicates that centralization offers greater efficiency and control and decentralization offers more flexibility to users. Unfortunately, these objectives are often in direct conflict. The resolution of this conflict is often determined by organizational politics and the relative strengths of the various users and the corporate computing staff. Preferably, the organization should properly consider such issues as equipment costs, effective utilization of computer personnel, and user satisfaction.

In actuality, many of the unique advantages of either approach have been reduced with the advent of new technologies for, and approaches to, computer-based information systems. For example, low-cost minicomputers have greatly offset the economy-of-scale argument. Techniques such as structured programming, top-down development, and chief programmer teams have reduced the impact of turnover. In other words, technology and personnel are not quite the deciding issues they were previously. Both centralized and decentralized approaches can be and are successfully used in organizations.

However, the fact that either approach can be used does not diminish the importance of careful evaluation and selection of an approach. Rather, an organization should consider its situation and consciously decide what approach is appropriate. Technology and personnel are supportive of either approach or of a combination of the two.

The following guidelines are helpful in making a decision on whether to centralize or to decentralize:

1. Centralized equipment and/or staff tend to be best suited for the following situations:
   a. Information systems for top-level management.
   b. Organization-wide functions such as payroll, personnel, and common accounting functions.
   c. Work that does not require rapid response time and that can be done more economically on a centralized basis.
   d. Supporting organizational functions that are too small to justify their own computing equipment and/or staff.
   e. Where integration of information technically requires centralized processing (e.g., online airline reservations).
2. Decentralized equipment and/or staff tend to be best suited for the following situations:
   a. Information systems for which there is no compelling reason to centralize (e.g., a stand-alone inventory system).
   b. Where rapid and flexible response time for development and/or production is required.
   c. Where minimizing the impact of computer downtime is essential (i.e., with only one centralized computer, a system failure can be a problem).

An underlying concept of the preceding guidelines is: Computing resources should only be centralized when economies and/or control considerations are compelling enough to compromise user convenience.

Organizational Location

The location of the data processing or information systems function is also an issue of debate and discussion. As conveyed in Table 9-1, most computing efforts are initially located in the department where the initial computer-based information systems originate. This is usually finance and/or accounting. However, the proliferation of computer applications in other functional areas of the organization generates interest in both a higher status and a broader orientation for the computing effort. Naturally, political as well as practical issues play a factor as the computing effort "percolates" up the organization.

Three basic approaches are used in positioning the computing effort.[7] Its possible locations are shown in Figure 9-2 and discussed below.

Location One: Under Department of Initial Applications

The computing effort is located in the finance and/or accounting department in a majority of organizations. This positioning is due to the early use of computers for financial applications. In organizations involved in large-scale scientific or engineering projects, the computer may be originally sponsored by the research or engineering department.

As long as computer processing is isolated to the department in which the computer is situated, this location is satisfactory. However, this isolation seldom exists for long; the computer's potential for other applications tends to proliferate quickly.

As computer applications develop outside the department in which the computer is originally located, the following problems may emerge:

1. **Lack of objectivity in setting priorities.** Computer personnel may tend to cater to the requirements of the department in which they work rather than to those of other departments.

---

7. Donald Sanders, *Computers and Management*, 1974, pp. 318–321.

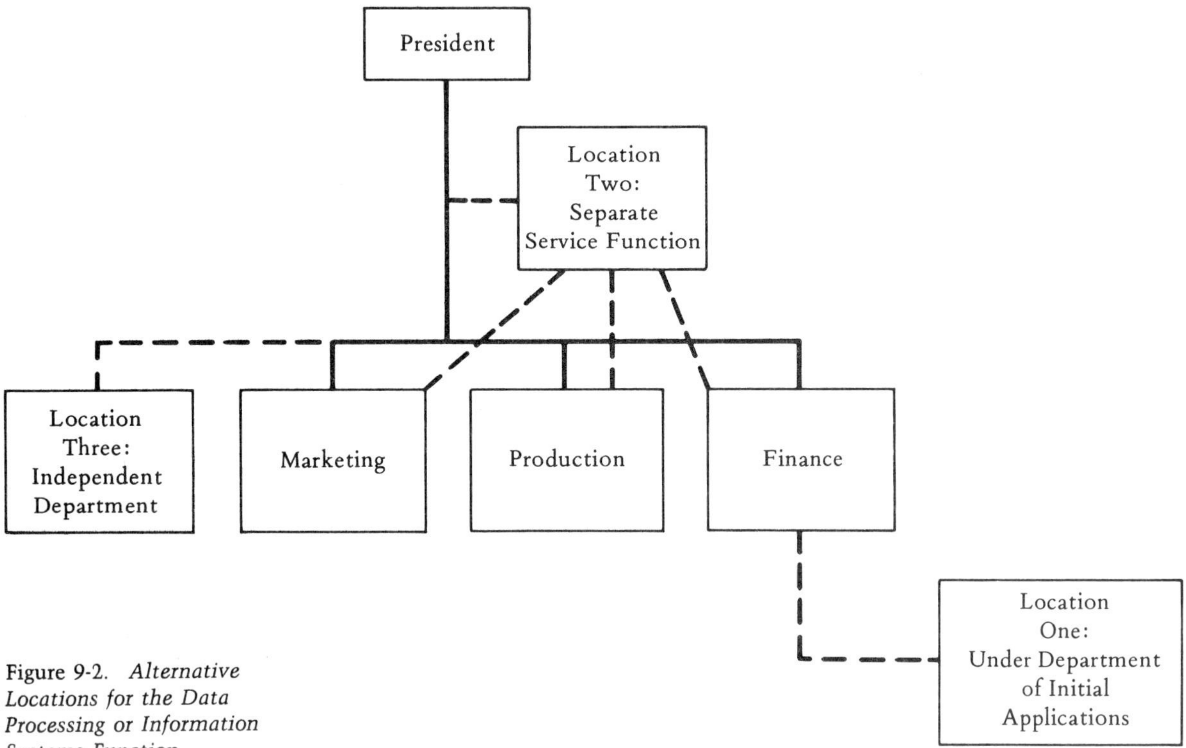

Figure 9-2. *Alternative Locations for the Data Processing or Information Systems Function*

2. **Limited vantage point.** Computer personnel may tend to view the organization in terms of the department in which they are located. If so, they may be unaware of problems or opportunities elsewhere in the organization.
3. **Lack of organizational status.** Organizational status and authority are lacking when the chief computer executive reports to a location several echelons down in one functional area of the organization. It is, therefore, difficult for him or her to influence top management in such areas as systems integration and interdepartmental information requirements.

Location Two: Separate Service Function

Problems of objectivity in priority setting can be overcome by setting up the computing effort as a separate service center to support the overall organizational computing requirements. The computing effort may report to a top-level executive or to an executive committee.

In this location, the computing effort occupies a position that is peripheral to the main organizational structure. The chief computing executive generally has little status or authority outside of his or her direct responsibilities. This can result in an unwieldy situation as he or she attempts to allocate what are usually scarce computing resources among executives who have high status and authority.

Location Three: Independent Department

A majority of information processing authorities contend that, to fully realize the potential of computer technology, the computing effort should be established as a separate, independent department. Their major claims are:

1. This location reflects the organization-wide scope of information because it does not isolate the computing effort in one functional area.
2. This location provides organizational status because it establishes the computing effort as a major organizational function.
3. This location encourages innovation and integration because it allows the computing effort to freely interact with other organizational functions.

Factors such as organizational size, complexity, the extent of computer usage, and managerial politics and personalities make it impossible to single out any one of the three locations as preferable for all organizations. However, location three is generally regarded as appropriate for medium to large organizations desiring to advance through the third and fourth stages of EDP growth.

PLANNING AND CONTROL

As shown in Table 9-1, planning and control of information systems development usually becomes a major concern during the third stage of EDP growth. However, recognizing the need for planning and control is only the first step. Determining what specific techniques are required to plan and control this often disruptive technology presents new challenges.

Project selection,[8] project management,[9] and auditing are the major concepts pertinent to management and control of computer-based information systems. These concepts are discussed below.

---

8. K. Eric Knutsen and Richard Nolan, "On Cost/Benefit of Computer-Based Systems," *Managing the Data Resource,* 1974, pp. 277–292.

9. F. Warren McFarland, "Effective EDP Project Management," *Managing the Data Resource,* 1974, pp. 273–307.

Project Selection

Traditionally, most organizational projects are selected in terms of their costs and benefits to the organization. At any point in time, several projects are usually worthy of consideration. Only a few can be active during a particular period of time. Usually, only projects offering highly attractive cost/benefit ratios are selected. Examples of commonly used techniques for evaluating cost/benefit are:

- Return on Investment (ROI)—Net project savings or revenue are divided by project costs to compute percentage return.
- Break-Even Analysis—Cash inflows from revenues (benefits) are matched against cash outflows for costs until they are equal (i.e., at a break-even point).

To conduct either of the preceding analyses requires that the costs and benefits be quantifiable and estimatable. Unfortunately, information systems projects do not lend themselves to easy quantification and estimation of costs, savings, or revenue that can then be cranked through ROI, break-even, or similar evaluations. The costs and benefits of computer-based information systems are varied, complex, interrelated, and difficult to quantify and estimate.

Costs and Benefits

Costs of information systems are especially difficult to estimate if the project is new and cannot be easily related to previous experiences. During the initial stages of systems development, cost estimates are commonly off by over 100 percent. For a typical system, it is not until after the system has been designed and the technology and personnel selected that there is enough project definition to approach an accurate estimate of the overall project costs.

Preliminary cost estimates should be considered tentative. The fact that they are tentative and subject to substantial variations reduces their value for quantitative analysis.

Benefits of information systems projects are more difficult to assess than costs are. Many qualitative and intangible benefits do not lend themselves to quantification or measurement.

Guidelines

Top-level management is familiar with and well equipped to deal with estimated and qualitative variables. Such evaluations are common to their organizational roles. Therefore, it is important to involve top-level management in the selection of projects. Their participation is best accomplished through the use of an executive steering committee that includes top-level management representatives from the major functional areas of the organization. This not only provides the unique vantage point of top management but also insures the necessary broad perspective.

The steering committee can review project proposals in both quantitative and qualitative terms. Project developers can present (in quantitative form) the costs and benefits that are readily measurable and easily understood. They can present short descriptive statements of benefits that cannot be readily quantified. Management personnel with experience and training can provide insight and give careful, broad thought to qualitative issues. This approach to project selection has been quite successful.

Project Management

After projects have been selected, they must be properly managed through the development cycle. The contingency view of management discussed in Chapter 3 states that the appropriate management approach is a function of the type of organizational process being managed. Information systems projects have varying properties that should be considered when choosing the management approach to use for a particular project.

Properties of Projects

The project properties that most significantly influence the overall nature of a project are:

1. **Predefined structure.** The more predefined structure a project has, the more easily it can be planned and controlled. For example, transaction processing applications inherently have a great deal of predefined structure. Their structure reduces the difficulty of designing computer applications to process them. Conversely, decision support systems (e.g., a market forecast system) are not usually well structured. They require considerable definition and structuring before they can be computerized.
2. **Stability of technology.** The greater the experience with a given technology to be used for a new system, the more predictable the systems development process. On the other hand, when a new information system is to use new and unproven

hardware and/or software, many unforeseen problems that impede the development process may arise.

3. **Size.** There is an inverse relationship between project size in terms of person years and costs, and the ability to accurately plan the number of person years and the costs that will be required to complete the project. That is, the larger the project, the more difficult to estimate the resources required to complete it.

Any given project can possess any variation of each of the preceding properties. For example, a project can have predefined structure, but use unstable technology and be a massive undertaking (e.g., most initial online airline reservation systems could be described in this fashion).

The more defined a project's structure is, the more stable the technology to be used is; and the smaller the size of the project, the more straightforward the project is. The more one or more of these properties deviates in the opposite direction, the less straightforward the project.

Techniques for Project Management

The techniques for managing information systems can be categorized as informal and formal. The orientation appropriate for a given project is contingent upon the properties of the project. Informal techniques are appropriate for projects that are not straightforward and predictable. Formal techniques are better suited for projects that are relatively straightforward and predictable.

Informal Project Management

Informal project management techniques allow projects to evolve as they are developed. Rather than try to fully predict all project requirements and problems in advance, management allows the project to progress at its own pace. Milestones or checkpoints are established to allow periodic review of that progress. These periodic evaluations allow management to determine if a project appears to merit further commitment of resources, if project adjustments are necessary, or if the project should be discontinued.

Formal Project Management

For projects that are more straightforward and predictable, more formalized management and control procedures are applicable.

The most commonly used formal project management technique is **PERT** (**P**rogram **E**valuation and **R**eview **T**echnique).[10] A PERT plan diagrammatically represents the tasks required to complete a project. It explicitly establishes sequential dependencies and relationships among the tasks. A PERT diagram consists of both activities and events. *Activities* are defined as time- and resource-consuming efforts required to complete a segment of the total project. *Events* represent the completion of segments or parts of segments of the project. Activities are represented by solid lines with directional arrows; events are represented by circles. Dotted lines are used to represent sequential dependencies where there is no activity between two events. That is, one event must precede another, but no task has to be performed to progress from the first event to the second one. Activities and events are coded or described to designate their functions in the overall project.

Figure 9-3 shows a PERT chart and an accompanying project table that defines the responsible personnel and estimated and actual times and costs for each task in the project. By comparing the actual times and costs with the planned times and costs in the PERT chart, management can monitor and control project performance.

A final advantage to PERT is that the total time required to complete the project can be determined by locating the longest path (in terms of time) in the chart. This path is referred to as the *critical path.* For scheduling purposes, any delay of tasks in the critical path is an equivalent delay on the overall project.

Guidelines

The inappropriate application of formal or informal project management techniques can have unfortunate consequences. The use of informal techniques for a straightforward project needlessly forgoes planning definition and control. This is unfair to the organization and to the systems developers whose performance cannot be evaluated as accurately as it could be if more formal techniques were used.

The use of formal techniques for projects that are not straightforward or predictable generally results in dysfunctional constraints on what needs to be a relatively innovative and creative process. Systems developers are often forced to cut corners and stifle innovative processes in order to keep the project on schedule and within budget. There must be sufficient slack to

10. For a more thorough treatment of PERT, see "Project Management with CPM and PERT" by Joseph J. Moder and Cecil R. Phillips, 1970, pp. 20–36.

Figure 9-3. A PERT Chart and Accompanying Project Table.

PERT Chart

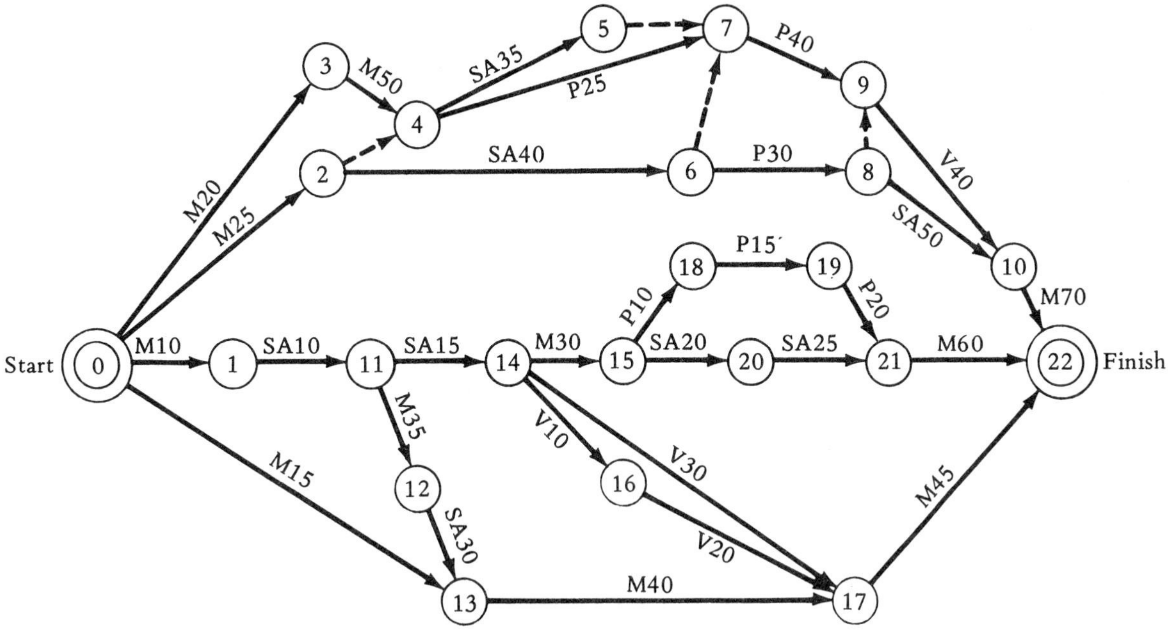

Project Table

Task	*Responsibility*	*Planned Start*	*Planned Completion*	*Actual Start*	*Actual Completion*	*Planned Costs*	*Actual Costs*
M10	JCW	1/10	1/15	1/10	1/15	N/A	N/A
M15	SCM	1/10	1/30	1/10	1/15	N/A	N/A
M20	JCW	1/15	1/30	1/15	1/30	N/A	N/A
M25	JCW	1/15	1/30	1/15	2/02	N/A	N/A
SA10	NJW	1/15	1/30	1/15	1/30	$ 800	$ 670
M50	SCM	1/30	2/05	1/30	2/04	N/A	N/A
SA15	NJW	1/30	2/07	1/30	2/06	$1,200	$1,250
SA40	BWB	1/30	2/25	2/02	2/26	$3,600	$3,200
SA35	RER	2/05	2/20	2/08	2/20	$ 500	$ 750
P20	JLP	10/01	10/25	10/05	10/25	$1,100	$1,400
SA50	NJW	10/25	11/10	10/20	11/10	$1,500	$1,500
V40	IBM	10/25	11/10			$5,000	
M60	JCW	10/25	11/20			$2,000	
M45	SCM	10/25	11/20			$1,000	
M70	JCW	11/10	11/20			$2,500	

Legend:

M = *Managerial* *SA* = *Systems Analysis* *P* = *Programming* *V* = *Vendor*

nurture innovation when approaching new areas of systems development.

The techniques used for projects can be combinations of formal and informal techniques. For example, if the specific project tasks and their sequential dependencies are known, but how long they will take and how much they will cost are not known, a PERT chart can be constructed without time and cost estimates. Such an approach provides more definition than a milestone approach without unduly constraining the project with time and cost estimates that may be unrealistic.

Mythical Person-Month

A key concept of project management is the defining and committing of human resources to projects. These resource commitments are generally expressed in terms of time periods such as person-months (or person-hours, -weeks, or -years).

A common management error is to assume that persons and person-months (or whatever time period is used) are directly interchangeable.[11] For example, a project requiring four people eight person-months (i.e., two months elapsed time) to complete usually cannot be completed in one month by adding four more people to the project. The complex interrelationships of most projects result in diminishing returns when more than a workable number of personnel are assigned to a project. If too many personnel are assigned to a project, they can literally get in one another's way to such an extent that the additional personnel are more of a hindrance than an asset to project completion.

The relationship of adding personnel to a project to the time required to complete the project is shown in Figure 9-4. Note that adding personnel expedites project completion up to a point, but then it begins to delay the project.

Operations Management

Once developed and implemented, information systems require ongoing planning and control. Planning continues in the form of scheduling of production processing and planning for systems maintenance as modifications or enhancements are required. Ultimately, planning involves the phasing out of obsolete systems as they are replaced by new ones.

---

11. Frederick P. Brooks, Jr., *The Mythical Man Month–Essays on Software Engineering*, 1974.

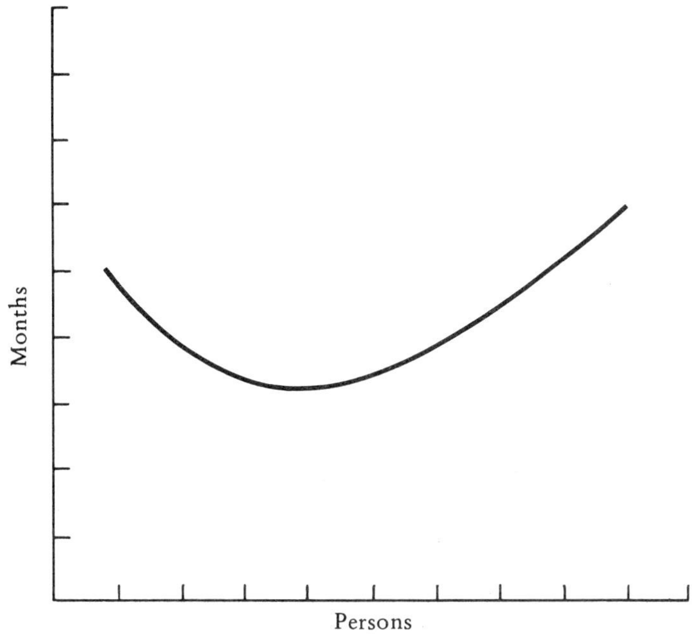

Figure 9-4. *Relationship of Adding Personnel to Time Required for Completion of a Project with Complex Interrelationships*

Control of information systems is required to insure that processing is conducted according to established procedures. It is also required to insure that the integrity of the systems documentation is maintained so that the system does not become dependent on a few key personnel for operation.

Audit Trail

The most comprehensive technique to insure that both users of systems and systems themselves are functioning properly is the use of an audit trail. To provide an audit trail, documentation of the when, what, and who of data modification is performed as a normal by-product of updating files. With an audit trail, data are captured and reported about: when the updated transaction occurred, what the data looked like before and after the transaction was executed, and who created the transaction. This is similar to the approach used in debugging systems as discussed in Chapter 8.

Audit trails can be maintained easily in batch systems. Transactions are recorded on documents and punched into cards; therefore, physical records of transactions exist after updates are made. Audit trails in online systems are more difficult to establish. Individuals can use terminals to modify data without creating recorded documents or punched cards. A common approach to address this problem is to require all terminal users to identify themselves (e.g., by use of an account number or a password). Transaction logs are kept, usually on disk. They document the terminal, the user, when data are changed, and what data are changed.

An audit trail can be used to spot-check transactions and to track irregularities. Also, the fact that audit trails exist serves as a deterrent to most unauthorized or improper use of systems.

Documentation Control

To insure that systems integrity is being maintained, periodic checks of the systems documentation should be made to verify that it is complete, current, and sufficient for systems operation in the event of turnover of key personnel.

The major areas of systems documentation are:

1. **Systems specifications.** The systems specifications are usually developed as a by-product of the systems design process. The systems specifications techniques discussed in Chapter 6 provide such documentation. However, care must be taken to insure that the documentation remains current and reflects any modifications (e.g., new reports, new data elements, or changes in validation rules) that occur after systems development is complete.
2. **Program documentation.** Program documentation usually consists of source listings, descriptive overviews of programs, and program flowcharts. When meaningful data names are used and structured programming is adhered to, the programs are easy to follow when they must be understood by programmers other than the original coders. It is, of course, important to insure that all program documentation reflects the latest modifications.
3. **Operating procedures.** Operating procedures describe the steps necessary to set up job control, tell how to mount special forms, indicate what files to use, and so forth. This documentation is usually straightforward, but it must be explicit so that processing is not dependent on one computer operator who has learned by experience or received additional verbal instructions.
4. **User manuals.** User manuals describe how to prepare transactions, request reports, interpret reports, and the like. They are usually based on systems specifications, restated in a form understandable to users.
5. **File documentation.** File documentation defines the tapes or disks that files are stored on. It also documents any file labels

and file retention periods. File documentation should be updated after each computer run or after each processing cycle to insure that current files will be used for subsequent processing.

The specific format of systems documentation is usually defined by the data processing or information systems department. Whatever the format, an organization should have backup copies of all documentation for recovery purposes in case the originals are destroyed or sabotaged.

There are two basic, but not mutually exclusive, ways to validate documentation. The first is to periodically review documentation and check it against current operations for inconsistencies. The second approach is to insure that vacation and/or rotating job assignments occur. This approach forces different personnel to operate different systems with only available documentation to go by. It more rigorously tests the quality of systems documentation, because it more definitely insures that systems operation is not personnel dependent.

Controlling Costs

The increasing costs of computing efforts generally encourage management to take cost-reducing measures during the third stage of EDP growth. (See Table 9-1 and Figure 9-1.) During the first two stages of EDP growth, computing costs are usually treated as organizational overhead; little or no information systems costs are absorbed by users. Since users are not financially affected by increases in demands for computer-based information services, they are not likely to be overly cost conscious in making additional requests for services. This can contribute to proliferating costs.

During the third stage of EDP growth, management frequently considers and implements chargeout systems that cause user departments to bear the expenses of the development and/or operation of their respective information systems. Research studies indicate that such chargeout systems do in fact cause users to be more cost conscious, resulting in better planning and control of information systems.[12,13] Also, users exhibit more positive attitudes about the information services they receive when they are

---

12. Richard L. Nolan, "Effects of Chargeout on User/Manager Attitudes," *Communications of the ACM*, Mar 1977, pp. 177–185.

13. Richard L. Nolan, "Controlling the Cost of Data Services," *Harvard Business Review*, July–Aug 1977, pp. 114–124.

paying for them and have more control over them (i.e., they can choose to buy or not to buy).

Other research indicates that the objectives of controlling costs and improving user attitudes can be accomplished by other means.[14] For example, a top-management steering committee may be used to allocate computing resources. When users know that all projects are reviewed and approved by top management, the users tend to exhibit the same conscientious behavior they show when they pay for services. This approach has the advantage of insuring better coordination and integration of information systems projects, due to the coordination and integrative efforts of top management. Such coordination and integration can be more difficult to achieve when users are functioning as independent "purchasing agents" of information.

The fact that each approach achieves both controlled costs and improved user attitudes indicates that the key issue in controlling costs and improving user attitudes is to involve top management and to make users more cost conscious.

BEHAVIORAL CONSIDERATIONS

The major orientation of the material in this book is toward the technical area of information systems. However, serious problems in the development and operation of information systems may result from the behavioral reactions of those affected by changes brought about by these information systems. An understanding of the nature and causes of such behavior is helpful to both management and systems analysts. They are in a better position to properly address resultant problems. Research by Gary W. Dickson and John K. Simmon provides a useful framework for understanding and minimizing dysfunctional behavioral reactions that may be brought on by information systems development and operation.[15] The following discussion is based upon their research findings.

Behavioral Groups

Four basic behavioral groups are involved and/or affected by information systems development and operation. These groups are described below.

---

14. James C. Wetherbe, "A Zero-Based Approach to Allocating MIS Resources," *Proceedings of the Ninth Annual Conference of the Society for Management Information Systems*, Sept 1977.

15. Gary W. Dickson and John K. Simmon, "The Behavioral Side of MIS, Some Aspects of the People Problem," 1970.

1. Operating personnel—Includes all nonmanagerial personnel in the organization except those classified as technical staff. Operating personnel are further subdivided into nonclerical and clerical. Nonclerical personnel provide the inputs to the system (sales, orders, time cards, production order, etc.). Clerical personnel process the inputs and convert them into forms suitable for processing.
2. Operating management—Includes personnel in positions ranging from first-line supervisor up to and including middle management.
3. Technical staff—Includes those most involved with the development and technical support of the system. Systems analysts and programmers are in this group.
4. Top management—Includes the chief executive officers of the organization. The president and vice presidents are in this group.

Dysfunctional Reactions

The most common forms of dysfunctional reaction to information systems are explained below.

1. **Aggression.** This behavior represents attacks (either physical or nonphysical) intended to disrupt or destroy the system. The most dramatic aggression behavior is sabotage. Less dramatic and more common behavior occurs in the form of attempts to "beat the system" or to disrupt it by outsmarting it.
2. **Projection.** This behavior represents attempts to blame the system for difficulties or problems that are, in fact, not caused by the system. When a problem occurs, personnel simply claim that the problem is a computer error or problem.
3. **Avoidance.** This behavior represents defensive reactions by personnel who withdraw from or ignore the system because they are intimidated, frustrated, or simply reluctant to familiarize themselves with the new system and any changes it has brought about.

During the development and operation of a new information system, all four of the behavioral groups usually experience varying degrees of uncertainty, unfamiliarity, and increased complexity. Because each group has a somewhat different relationship with information systems, the groups are also affected in different ways by the new system. Usually, they exhibit different behavioral reactions. These reactions are discussed below.

Operating Personnel

Clerical operating personnel are particularly affected by new information systems. Their jobs may be changed considerably or eliminated. Work and interpersonal relations are usually modified. There may be increased work rigidity and time pressure. The usual response to such effects is projection behavior: all of the undesirable effects and any other ensuing problems are blamed on the computer and its technicians. Though clerical personnel may want to indulge in avoidance behavior, their low-level positions in the organization tend to make that impractical.

Nonclerical operating personnel are less affected by new information systems. Usually, they are only required to fill out new or additional forms to provide input to the new system. They tend to demonstrate aggression behavior because of the increased inconvenience imposed on their daily work activities.

Operating Management

Operating management is usually highly affected by new information systems. As a result of the additional information available from new systems, top management is usually more informed about the effectiveness and efficiency of operating managers. This often leads to closer evaluation and control of their activities.

Many of the decisions made by operating managers may be displaced by programmed decision making performed by the new information system. Programmed decisions usually create free time for managers. This increased free time allows operating managers to engage in other, more creative and constructive efforts. However, the overall effect as perceived by the managers is a change in their job content. Operating management, therefore, has reason to feel their status or power is threatened. They often experience feelings of insecurity. These feelings may be strong. If so, reactions of aggression, projection, and avoidance are apt to occur.

Top Management

Top management is less affected by and concerned with new information systems than other groups are. They may feel some insecurity because a technology they do not understand plays a larger role in their organization. However, their high-level positions provide them the option of avoiding new systems. They have little reason to exhibit projection or aggression behavior.

Top management is usually more involved with information systems that do more than transaction processing, information providing, and programmed decision making. In other words, they are more affected by systems that provide decision support capabilities. However, decision support capabilities are not commonly included in many organization's information systems.

Technical Staff

Technical staff are the "change agents" whose careers are based and dependent upon developing new and better information systems. Therefore, they seldom demonstrate dysfunctional behavior toward new systems, other than perhaps occasionally overpromoting them.

Guidelines for Minimizing Dysfunctional Behavior

The recognition of the nature and causes of dysfunctional behavior as it relates to new information systems does not, by itself, solve any problems. The following discussion pertains to steps or procedures that can be used to minimize dysfunctional behavior. There is no single, best solution. Many questions about human behavior are yet to be answered. Accordingly, the following guidelines are plausible solutions but they are not absolute answers.

Top Management

A foremost guideline for overcoming dysfunctional behavior toward new information systems is to obtain top management involvement and support. The importance of such involvement and support in project selection, project management, operations management, and controlling costs is stressed earlier in this chapter. In spite of the importance of doing so, getting top management involved can be difficult. As discussed earlier, top management's likely reaction to new information systems is avoidance. The systems analyst must take care to describe information systems problems in nontechnical terms when communicating with top management and soliciting their support. Top-level managers, like everyone, are much more comfortable with things they understand and, consequently, are much more likely to get involved.

An additional means of gaining top management support is to stress that information systems projects are consistently more successful from an operational, economical, and technical perspective when top management is involved and supports them. Indeed, eliminating avoidance behavior in top management can significantly contribute to eliminating dysfunctional behavior in the remainder of the organization.

Operating Personnel

Nonclerical operating personnel generally react to new information systems with aggression behavior, invoked by what are usually increased requirements for input preparation. For example, salespersons may have to fill out new reports describing factors that cause the loss of any customer or client. To the salespersons, these reports may appear to be needless additional paperwork. However, information generated from the reports may be used to recognize patterns in the competitiveness of the company's product line. Such information can then be used to make decisions about product modifications or new product developments. These will, in turn, enhance the salespersons' abilities to regain lost customers and acquire new ones. If the salespersons understand and believe this strategy, they are less likely to react negatively to the additional input requirements.

The key issue is that if information is valuable, those providing it should be sold on the importance of collecting and processing it. This is best accomplished when top management communicates their support of the information system and its use. Top management is usually more effective than technical staff in gaining support for a new information system from operating personnel.

Clerical operating personnel's common dysfunctional reaction of projection is caused by fears that their jobs may be changed adversely or even eliminated. This reaction is often brought on because clerical personnel are the last to know what is going on as new systems are being developed. The almost secretive atmosphere that surrounds the new systems creates concerns that something bad is about to happen. Such concern can be minimized if top management provides an atmosphere of openness and security about forthcoming changes. When clerical personnel are made aware of what is taking place and are assured of their job security, projection behavior can be reduced. It is virtually impossible for the systems analyst to create such an atmosphere without top management's involvement and support.

Operating Management

As computer technology first began to be used by organizations, much was said and written about the displacement of operating management. It appeared to many that the ability of the computer to process large volumes of data and generate solutions would reduce or eliminate the need for operating management. Understandably, operating managers were, and often still are, threatened by the computer. It is not too surprising that operating managers often exhibit all three forms of dysfunctional behavior: aggression, projection, and avoidance.

The computer can (and does) make a large number of decisions previously made by operating managers. It is quite efficient at making programmed decisions for highly defined and structured problems. However, the computer has not displaced operating managers. Rather, these managers have been released to address much more complicated and less structured problems created by more complicated organizational processes. This pattern is apt to continue. Therefore, the need for operating managers, with even greater skills, still exists.

The fact remains, however, that operating managers continue to experience anxiety as new information systems are developed and put into operation. They are often required to accept changes in job content as the computer takes over routine tasks. To minimize dysfunctional behavior resulting from these circumstances, top management should be open about forthcoming changes and include operating management in the design of new information systems. Only top management can convincingly stress that operating managers will play a more challenging and important role in the organization, even though some of their decision-making activities will be relinquished to the computer.

General Guidelines

The preceding discussion addresses guidelines for minimizing dysfunctional behavior within each behavioral group. A few guidelines for minimizing dysfunctional behavior apply to all behavioral groups. They are listed below.

1. Information systems should be thoroughly tested and validated prior to implementation. A malfunctioning system can generate undesirable behavior on the part of all personnel.
2. The development and implementation schedule should be realistic. If personnel are rushed into a new system, problems and negative reactions are likely to occur.
3. Systems clarity should be strived for by keeping systems as simple as possible, providing adequate training, and designing systems outputs to fit the needs of users. Lack of systems clarity encourages personnel to keep their own (private) man-

ual information systems because they are afraid to use or depend on the computer-based information systems.

4. Any changes in job content resulting from new information systems should be recognized and adjusted for. In particular, job performance evaluations and the accompanying reward systems should be modified to reflect job changes. For example, if salespersons are required to fill out new competitive analysis forms, the manner in which they perform this task should be evaluated and properly considered as part of their overall job performance. Alternatively, if a part of an employee's job is eliminated, he or she should no longer be evaluated on that basis. For example, if the computer begins to make inventory reorder decisions, the inventory manager should not be held responsible for inventory decisions that he or she cannot control.

As a final note, the importance of allowing those effected by change to participate in bringing about that change cannot be overemphasized. It is often thought that most people are simply reluctant to accept change. However, this belief is contradicted by the often enthusiastic manner in which people accept changes in such things as clothes and hair fashion. People enjoy driving the latest car or having the latest gadgetry (e.g., electronic calculators and microwave ovens). The reason for this apparent contradiction is that people, as consumers, have a great deal of control over these changes. They can decide by their purchasing behavior if they want to accept or reject new products. If these products were forced on people, they would likely react quite differently. By allowing organizational personnel to participate in organizational changes, acceptance and even enthusiasm is much more probable.

SUMMARY

The results of computer-based information systems in most organizations have fallen substantially below management's expectations. This has been attributable to a great extent to the lack of and difficulty of proper planning and control of the computing effort. Computing efforts in organizations go through four basic stages of growth: (1) initiation, (2) expansion, (3) formalization, and (4) maturity. During the third stage, management develops sufficient concern and expertise to apply proper planning and control techniques to more efficiently and effectively direct the computing effort.

The organizational location of the data processing or information systems function is a key planning and control issue. The degrees of centralization and decentralization of computer operations, systems analysis, and programming are variables that should be resolved on the basis of the needs and requirements of the organization. The computing effort should be located within the organization such that it can properly service all information systems users and have sufficient organizational status to properly influence the coordination and integration of information systems development.

Planning and control of the information systems is concerned with project selection, project management, operations management, and controlling costs. In performing each of these functions, management should consider the properties of projects properly in terms of their visible and hidden costs and benefits; their degree of structure, technological stability, and size; the need to provide processing and documentation control; and the need to control costs. Accomplishing these functions is not a trivial effort. Above all, it requires a concerned, involved, and supportive top management.

Behavioral considerations associated with information systems development are extremely important but frequently overlooked. Dysfunctional behavior resulting from poorly managed systems implementation can be disruptive and even destructive to the success of information systems. The four behavioral groups affected by information systems include operating personnel, operating management, technical staff, and top management. The different groups react to new information systems by exhibiting one or more behavioral reactions of aggression, projection, or avoidance. Dysfunctional behavior can be minimized by properly considering why it occurs and taking appropriate managerial action.

EXERCISES

1. Discuss why a systems analyst should have an understanding of the proper techniques of administrating information systems projects.
2. What are the reasons for the lack of and difficulty of planning and control of information systems projects?
3. Explain the four stages of EDP growth and their significance to the planning and control of computing efforts in organizations.
4. Why can't most information systems be selected on a simple cost/

benefit basis? What additional factors must be considered? How can they be considered?

5. Discuss the advantages and disadvantages of centralization and decentralization of the data processing or information systems function.

6. What are the basic organizational locations for the data processing or information systems function? What are the considerations in selecting a location?

7. Discuss the major properties of information systems projects and how they affect the manner in which projects should be managed.

8. Explain the meaning of the "mythical person-month" and the problems related to it.

9. Identify the major behavioral groups affected by an information system and their usual forms of dysfunctional reactions to new information systems.

SELECTED REFERENCES

Brooks, Frederick P., Jr. *The Mythical Man Month—Essays on Software Engineering*. Reading, Mass.: Addison-Wesley, 1974.

Business Week Editorial, "Business Takes a Second Look at Computers," *Business Week*, June 5, 1971, pp. 59–136.

Dearden, John, and Richard Nolan. "How to Control the Computer Resource," *Harvard Business Review*, Nov–Dec 1973, pp. 68–78.

Dickson, G. W., and John K. Simmon. "The Behavioral Side of MIS, Some Aspects of the People Problem," Bloomington, Ind.: Indiana University Press, 1970.

Drucker, Peter F. *Managing for Results.* New York: Harper & Row, 1964.

Glaser, George. "The Centralization vs. Decentralization Issue: Arguments, Alternatives, and Guidelines," *Data Base* Vol. 2, No. 3 (Fall/Winter 1970), pp. 1–7.

Golub, Harvey. "Organizing Information System Resources: Centralization vs. Decentralization," from *The Information Systems Handbook.* ed. McFarlan and Nolan. Homewood, Ill.: Dow-Jones-Irwin, 1975.

Guthrie, A. *Attitudes of Middle Managers toward Management Information Systems.* Ph.D. Diss. Seattle: University of Washington, 1971, pp. 61–67.

Head, Robert V. "The Information Systems Manager as the Administrator of a Major Corporate Function," from *The Information Systems Handbook.* ed. McFarlan and Nolan. Homewood, Ill.: Dow-Jones-Irwin, 1975, pp. 61–64.

Knutsen, K. Eric, and Richard Nolan. "On Cost/Benefit of Computer-Based Systems," *Managing the Data Resource.* ed. Nolan. St. Paul, Minn.: West, 1974, pp. 277–292.

Lawrence, Paul, and Jay Lorsch. *Organization and Environment—Managing Integration and Differentiation.* Boston: Harvard Business School Division of Research, 1969.

Leavitt, H. "Applied Organizational Change in Industry: Structural, Technological, and Humanistic Approaches," in *Handbook of Organizations.* Chicago: Rand McNally, Mar 1965, pp. 1144–1170.

McFarland, F. Warren. "Effective EDP Project Management," *Managing the Data Resource.* ed. Nolan. St. Paul, Minn.: West, 1974, pp. 273–307.

McFarland, F. Warren. "Management Audit of the EDP Department," *Harvard Business Review,* May–June 1973, pp. 131–142.

McKinsey & Company, Inc. "Unlocking the Computer's Profit Potential," *The McKinsey Quarterly,* Fall 1968, pp. 17–31.

Miller, James C. "Corporate Organization and Information Systems," from *The Information Systems Handbook.* ed. McFarlan and Nolan. Homewood, Ill.: Dow-Jones-Irwin, 1975, pp. 32–60.

Moder, Joseph J., and Cecil R. Phillips. "Project Management with CPM and PERT," New York: Litton, 1970, pp. 20–36.

Mulvihill, Dennis E., and Burton J. Cohen. "Strategy Formulation and Information Systems: Setting Objectives," from *The Information Systems Handbook.* ed. McFarlan and Nolan. Homewood, Ill.: Dow-Jones-Irwin, 1975, pp. 19–31.

Nolan, Richard L. "Controlling the Cost of Data Services," *Harvard Business Review,* July–August 1977, pp. 114–124.

Nolan, Richard L. "Effects of Chargeout on User/Manager Attitudes," *Communications of the ACM* Vol. 20, No. 3 (Mar 1977), pp. 177–185.

Nolan, Richard L. "Managing the Computer Resource: A Stage Hypothesis," *Communications of the ACM* Vol. 16, No. 7 (July 1973), pp. 399–405.

Nolan, Richard L. "Plight of the EDP Manager," *Harvard Business Review,* May–June 1973, pp. 143–152.

Nolan, Richard L., and Cyrus F. Gibson. "Managing the Four Stages of EDP Growth," *Harvard Business Review,* Jan–Feb 1974, pp. 76–88.

Porter, L. W. "A Study of Perceived Need Satisfaction in Bottom and Middle Management Jobs," *Journal of Applied Psychology* Vol. 45, No. 1 (Feb 1961), pp. 1–10.

Reichenbach, Robert, and Charles Tasso. *Organizing for Data Processing.* AMA Research Study 92. 1968, pp. 71–86.

Sanders, Donald. *Computers and Management.* New York: McGraw-Hill, 1974, pp. 318–321.

Wetherbe, James C. "A Zero-Based Approach to Allocating MIS Resources," in *Proceedings of the Ninth Annual Conference of the Society for Management Information Systems,* Sept 1977.

Wetherbe, James C., and Carlton J. Whitehead. "A Contingency View of Managing the Data Processing Organization," *Management Information Systems Quarterly* Vol. 1, No. 1 (Mar 1977), pp. 19–25.

SECTION

III Advanced Systems Analysis

In this final section of the text, two emerging trends in information systems development are discussed—database management systems and decision-support systems. These topics are discussed in Chapters 10 and 11, respectively. The last chapter in the text is a look at the future of systems analysis and information systems.

10 Management Information Systems and Database Management Systems

The use of database management systems represents one of the most significant trends in the field of computer-based information systems. A *database management system* (DBMS) is a collection of software for processing a collection of interrelated data, known as a *database.* In this chapter, both the need for and the characteristics of database technology are discussed. Then the effects of database management systems on the systems development process are pointed out.

MANAGEMENT INFORMATION SYSTEMS

One of the greatest expectations of computer-based information systems by management was that they would place massive amounts of information at management's immediate disposal. Many managers envisioned their offices as equipped with computer terminals that would allow them to monitor the operations of their organizations by the mere entering of a few commands into the terminals. All organizational data would be integrated into one single integrated database.

These expectations and their lack of fulfillment have produced one of the longest "expectation gaps" associated with computer technology. Shortly after the recognition that this initial perception of a management information system (MIS) would not be forthcoming, John Dearden wrote an article entitled "MIS Is a Mirage."[1] This article created a great deal of discussion and debate about whether or not MIS was a realistic concept.

As with many new concepts or ideas, the initial enthusiasm for MIS fostered inflated expectations and often gross misconceptions. More experience and research with MIS has provided a more realistic perspective of what role computer-based information systems are to play in organizations. Some of the original expectations for MIS have been or are being realized; others have not been and

1. John Dearden, "MIS Is a Mirage," *Harvard Business Review,* Jan–Feb 1972, pp. 90–99.

may never be realized. However, it is generally accepted that MIS is a meaningful and valid concept.

Two factors have inhibited or prevented the achievement of the original concept of MIS. These factors are based upon the characteristics of management's information requirements and the practicalities of integrating all organizational information. These topics are discussed below.

Characteristics of Management's Information Requirements

Prior to attempts to design comprehensive MIS's for management, there was somewhat of a dearth of information about what executives did or how they did it. However, as discussed in Chapter 3, it is now recognized that most of the decisions made by middle management and, particularly, by top management are (1) judgmental and subjective in nature, and (2) based upon information that is largely external to the organization and therefore usually incomplete.

The computer's ability to perform extremely fast retrieval of and/or calculations on highly structured data is of less applicability in the realm of top management. For example, though computers have been highly productive in processing sales orders and posting accounting transactions, they have been less successful at determining what actions competitors, customers, or the government may take.

The computer is, however, demonstrating increasing potential in the area of decision-support systems. This increased potential is a result of both a better understanding of management's decision-making process and enhanced sophistication of computer technology. For example, the computer can be used to simulate the results of different decisions, given different sets of events. Such a system allows management to play "what if" in considering different decisions. That is, the computer can simulate, with varying degrees of accuracy, what would happen in each case. Consequently, management can more accurately consider the likely results of several alternative decisions with respect to a given problem or opportunity, before actually selecting the course of action to be taken.

Decision-support systems are upgrading the use of computer technology from the operational levels of transaction processing and reporting to activities that directly support management. Decision-support systems offer great potential for achieving many of the unfulfilled expectations of MIS. A more extensive treatment of decision-support systems is presented in the next chapter.

Integration of Organizational Information

Though the difficulty of structuring and processing information for management has been one factor constraining the development of an MIS, a second, and often more perplexing, problem has been the inability to integrate organizational information. The need for integrated information for higher levels of management is best illustrated by the pyramidal structure of most organizations: there exists an ever-widening succession of levels from the top to the bottom of such organizations.

If an organization is to remain viable and operational, the activities of lower-level functional units must be coordinated and conflicts resolved by the integrative activities of higher levels of management. Ultimately, the chief executive officer must insure the integration of the activities of the overall organization. For example, production activities must be coordinated with marketing activities in order to maintain a balance between products produced and products sold. Coordination of these different organizational activities requires the integration of information about the activities such that they can be properly managed and controlled.

The speed and capabilities of computer systems appeared to offer great promise in overcoming the information integration constraint inherent with manual information systems. But managers of many organizations that have invested in computer technology have been quite disappointed; much of the information stored in their computer systems cannot be integrated easily.

A Banking Example

The reasons for the lack of integration of organizational information are illustrated by the following example. Consider a bank that installs its first computer system. The largest and most cost-effective application to computerize is demand deposit accounting (DDA), or checking accounts. A systems analyst is assigned to work with users to automate this system. DDA is a highly structured process that lends itself to computer processing. Account numbers can be used as keys, and the data to be stored in computer files is primarily the data used in the manual system, with some enhancements. Figure 10-1 depicts the data contained in a DDA

1. Account Number	17. Officer Code
2. Name	18. Date DDA Opened
3. Spouse's Name	19. Current Balance
4. Social Security Number	20. Last Statement Date
5. Telephone Number	21. Balance Forward—Last Statement
6. Date of Birth	22. Amount of Credits—Current Period
7. Sex	23. Number of Credits—Current Period
8. Marital Status	24. Amount of Debits—Current Period
9. Number of Children	25. Number of Debits—Current Period
10. Number of Dependents	26. Number of Returned Checks Year-to-Date
11. Rent or Own Home	27. Date of Last Deposit
12. Occupation Code	28. Service Charge
13. Years Employed	29. Date of Last Overdraft
14. Income Range	30. Amount of Last Overdraft
15. Credit Rating	31. Number of Overdrafts
16. Line of Credit	

Figure 10-1. *Contents of a DDA File*

file. Once stored in the computer system, the DDA file can be used to generate a variety of reports and/or terminal displays associated with the management and control of DDA.

During the development and implementation of the DDA system, the bank decides to begin implementation of a computer-based savings accounting system. Another systems analyst is assigned to work with users to automate this system. Input transactions, reports, and files are designed, based upon the manual savings accounting system, with several enhancements. The contents of a file designed for the savings accounting system are shown in Figure 10-2.

1. Account Number	17. Officer Code
2. Name	18. Date Savings Account Opened
3. Spouse's Name	19. Current Balance
4. Social Security Number	20. Last Statement Date
5. Telephone Number	21. Balance Forward—Last Statement
6. Date of Birth	22. Date of Last Deposit
7. Sex	23. Amount of Last Deposit
8. Marital Status	24. Low Balance—Current Period
9. Number of Children	25. Number of Withdrawals—Current Period
10. Number of Dependents	26. Amount of Credits—Last Statement
11. Rent or Own Home	27. Number of Credits—Last Statement
12. Occupation Code	28. Amount of Debits—Last Statement
13. Years Employed	29. Number of Debits—Last Statement
14. Income Range	30. Excess Activity Charges
15. Credit Rating	31. Anticipated Interest—Current Period
16. Line of Credit	32. Interest Earned Year-to-Date

Figure 10-2. *Contents of a Savings Accounting File*

1. Account Number
2. Name
3. Spouse's Name
4. Social Security Number
5. Telephone Number
6. Date of Birth
7. Sex
8. Marital Status
9. Number of Children
10. Number of Dependents
11. Rent or Own Home
12. Occupation Code
13. Years Employed
14. Income Range
15. Credit Rating
16. Line of Credit
17. Officer Code
18. Collateral
19. Amount of Payment
20. Mode of Payment
21. Number of Times Refinanced
22. Late Charge Indicator
23. Maturity Date
24. Number of Payments
25. Interest Rate
26. Original Amount of Loan
27. Life Insurance Premium
28. Balloon Payment
29. Date of Last Activity
30. Current Balance
31. Interest Earned Year-to-Date
32. Number of Payments to Date
33. Number of Late Notices Sent
34. Number of Returned Checks
35. Last Statement Date

Figure 10-3. *Contents of an Installment Loan File*

The bank continues its development of computer-based information systems by automating its installment loan system and its mortgage loan system. Existing account numbers are used for each system, and existing data capture methods are expanded to obtain all of the data shown in Figures 10-3 and 10-4.

1. Account Number
2. Name
3. Spouse's Name
4. Social Security Number
5. Telephone Number
6. Date of Birth
7. Sex
8. Marital Status
9. Number of Children
10. Number of Dependents
11. Rent or Own Home
12. Occupation Code
13. Years Employed
14. Income Range
15. Credit Rating
16. Line of Credit
17. Officer Code
18. Original Date of Loan
19. Property Identification
20. FHA or VA Number
21. Late Charge Code
22. Amount of Payment
23. Mode of Payment
24. Escrow Payment
25. Original Amount of Loan
26. Maturity Date
27. Interest Rate
28. Terms in Months
29. Date of Last Appraisal
30. Amount of Appraisal
31. Purchase Price
32. Percent of Loan Guaranteed
33. Late Charge Rate
34. Date of Last Activity
35. Escrow Balance
36. Number of Late Notices
37. Number of Months to Go
38. Interest Paid Year-to-Date
39. Principal Remaining

Figure 10-4. *Contents of a Mortgage Loan File*

1. Social Security Number	11. Vacation Leave
2. Name	12. Federal Withholding
3. Address	13. FICA
4. Department Code	14. State Taxes
5. Job Title	15. Insurance
6. Date of Hire	16. Regular Earnings Year-to-Date
7. Marital Status	17. Federal Withholding Year-to-Date
8. Number of Dependents	18. FICA Year-to-Date
9. Pay Rate	19. State Taxes Year-to-Date
10. Sick Leave	20. Net Earnings Year-to-Date

Figure 10-5. *Contents of the Payroll File*

The bank also automates its payroll system. The file for this system includes the data shown in Figure 10-5.

The development of these five information systems has spanned four years and cost several hundred thousand dollars. Bank employees and managers responsible for DDA have their own information system; bank employees and managers responsible for savings accounts have their own information system; and so forth. However, there are some serious problems. Each system was developed as an independent subsystem; no thought was given to interfacing the subsystems to provide integrated processing or reporting.

Redundancy

The first problem with the bank's information systems is that there is a great deal of redundancy in the data collected, stored, and processed. Consider a bank customer who has a checking account, a savings account, an installment loan, and a mortgage loan. Though the bank is dealing with one customer, its information systems treat this customer as four customers. The customer has four account numbers (one in each information system).

Certain data are redundantly collected, stored, and processed in each system. Specifically, data elements 1 through 16 are identical in each of four files (see Figures 10-1 through 10-4).[2] If a customer has more than one checking account, savings account, or loan, the redundancy becomes even greater. This redundancy creates the following problems:

2. These data elements were made identical for the purposes of this example. In actuality, the systems would share many common data elements, but files would tend to vary some in their names, sequence, and inclusion of elements.

1. Customers are required to supply much duplicate data for each account they open, even when some or all of the data needed has been collected previously.
2. Storage space is wasted because the same data are stored in different places in the same file and/or in different files.
3. Processing time is wasted. For example, a customer's address may, due to redundancy, be stored in eight different places in the bank's computer files. If the customer's address changes, all eight addresses must be updated to keep the data used in all information systems current.
4. Inconsistencies and/or other errors develop in data files. The majority of information systems fail to update all redundant data. For example, a customer's address may be updated in the DDA file but not in the savings accounting or loan files. Consequently, there are inconsistencies as to the address of that customer.

Integration

Besides creating inefficient redundancies, the independent subsystem structure causes difficulties in integrating information. The systems and files have been developed along departmental or functional boundaries. The account numbers are not logically related and cannot be used for cross-referencing customer's accounts. This seriously limits reporting capabilities. For example, a loan officer may want to check information pertaining to a loan applicant's checking and savings accounts. However, there is no linkage to these data from the loan system. Indeed, the loan officer may have to ask the loan applicant if he or she has checking and/or saving accounts with the bank and what his or her account numbers are.

Consider a case where the management of the bank wants to increase mortgage loans to offset several large savings deposits. Management decides to send letters encouraging specific customers to consider buying homes, using convenient financing available through the bank. Management also decides that the best customers to send such letters to are customers meeting the following criteria:

1. Customers who do not own homes but are married.
2. Customers who have good checking account records (i.e., few or no overdrafts).
3. Customers with sufficient funds in their savings accounts to make down payments on homes.

4. Customers who have good payment records on any installment loans with the bank.

Though the data necessary to identify such customers are available in the different files of the different information systems (see Figures 10-1 through 10-4), there is no convenient way to integrate them. Extensive programming and clerical work are required to satisfy such an information request. Management is understandably disappointed.

The problem is that the bank's information systems are not designed to integrate information to serve management's needs. Integration of information is more readily achievable when considered prior to developing and implementing systems. After the fact, integration can be unwieldy; it may even require new systems development.

It is important to point out that not all organizational information needs to be integrated. For example, the contents of the payroll file in Figure 10-5 are independent of the contents of other files. Though bank employees may have checking accounts, savings accounts, and/or loans with the bank, there is little incentive to endure the expense of structurally relating these data in the computer system (i.e., because a bank usually has several thousand accounts but only a few hundred employees). Any requirements to relate employee data to data in the other information systems can usually be handled more cost effectively on a manual basis.

Thus, organizational information may not be integrated due to oversight or because the data cannot be integrated practically. The former situation is caused by poor systems analysis and design. The latter is caused by the nature of the data being processed. Figure 10-6 shows the relationships of the data contained in the bank's five files.

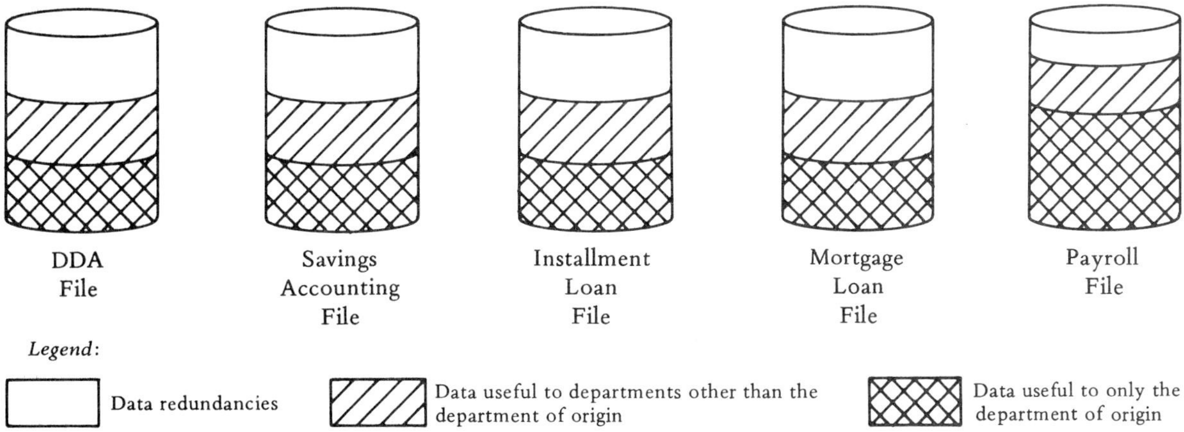

Figure 10-6. *The Relationships of Data in a Bank's Files*

The scenario of the bank can be translated into other organizational settings. For example, in a university, student data, course data, faculty data, and classroom facilities data should all be integrated. In a manufacturing plant, sales data, inventory data, production resources data, and purchasing data should be integrated.

The Failure to Integrate

In retrospect, it is not difficult to see that, when appropriate, information systems should be integrated. However, few organizations have attempted such integration in their initial attempts to develop computer-based information systems. The failure to approach information systems in an integrated fashion illustrates a key concept discussed in Chapter 5: the difference between *reacting to problems* and *recognizing opportunities.* Many information systems are developed as high-speed automated versions of existing manual systems. Since manual systems do not lend themselves to integration with other manual systems, the computer versions of these systems retain this nonintegration orientation. The opportunity to integrate the manual systems as they are computerized is not exploited. Rather, the personnel developing the systems react to management's demands for integrated information after it becomes apparent that there is a structural deficiency within their initial development efforts.

As absurd as it often seems, the reactive approach to problem solving is an all too common phenomenon in our industrialized world. Consider the energy crisis. That there is not an infinite supply of oil resources should have been recognized long before there was an actual shortage. Energy alternatives should have been pursued some time ago.

Difficulties in Integrating Data

A number of organizations, have attempted to integrate their information systems either initially or during early revisions of their initial systems. Though their insight and efforts have enabled them to more readily integrate data and, therefore, reduce redundancies, they have still encountered problems. The programming of data relationships that transcend departmental or functional boundaries is extremely complex. To achieve integration requires consolidation of files and/or linking of data stored in separate files.

In our example above, four systems of the bank (DDA, savings accounting, installment loans, and mortgage loans) can be integrated into a single centralized file. One way this can be ac-

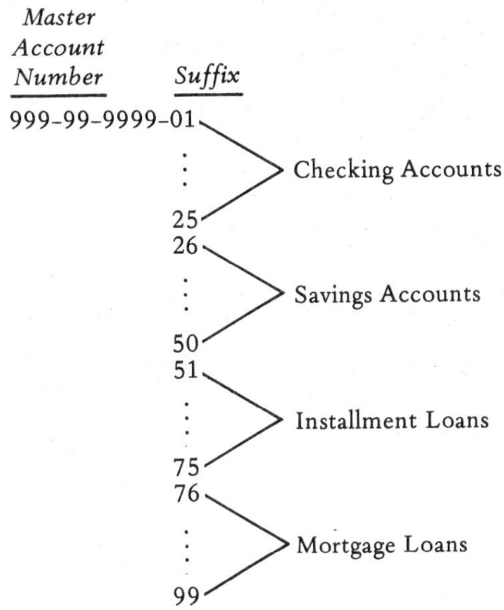

Figure 10-7. *Use of Suffixes to Achieve Centralized Account Numbers at a Bank*

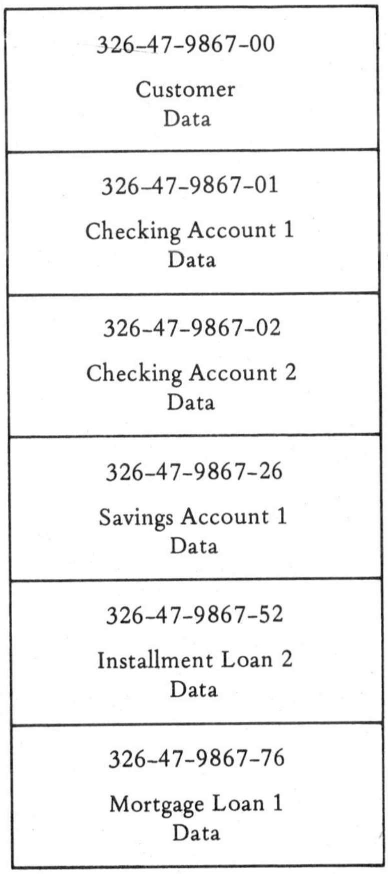

Figure 10-8. *A Customer Record from a Consolidated File of a Bank*

complished is by assigning to each customer a single account number and using suffixes to designate the accounts the customer has with the bank. Figure 10-7 illustrates such a customer account structure. Note that a customer can have several checking accounts, savings accounts, and loans. The unique identity provided for each customer account allows the accounts to be consolidated into one file as shown in Figure 10-8.

This bank is unlike most organizations in that, generally, an organization's separate files cannot practically be consolidated into a single file. Where consolidation is not possible, integration of data is sometimes accomplished by linking the data from separate files.

A University Example

Consider a student file and a class schedule file used in a registration system at a university. Integration of the data in the files is required to determine what classes each student is scheduled for and what students are scheduled for each class.

In a poorly designed registration system, integration of data can be accomplished by redundantly storing pertinent class data (class number, name, instructor, room, days, time, etc.) for each student in the student file. Also, all pertinent data about students scheduled (student number, name, major, classification, etc.) for each course must be redundantly stored in the class schedule file. These redundancies create problems. If 200 students are enrolled for a class, the data pertinent to that class are stored 200 times in the student file (i.e., once for each student). If the room number in which the class is to be held is changed, updating these data requires updating all 200 copies of it stored in the student file as well as the original version stored in the class schedule file.

Rather than redundantly store class data in the student file and student data in the course file, a more efficient approach can be taken. It involves storing only the keys to the class and student data. For example, the only class data that are stored in the student file are the class numbers of each class a student is scheduled for. The class number can then be used as an index to the class schedule file, which can then be accessed to obtain the data describing the class. If 200 students are enrolled in a course, the data for each of the students include an index to the class data (but no additional class data are actually stored in the student file). Consequently, class data such as time, place, and instructor need only be changed in the course file to properly reflect changes for each student scheduled for the class. A reciprocal arrangement is possible for the class schedule file. Only the index to the students scheduled for a class need be stored in the class schedule file—not the student data.

The concept of linking data between files is illustrated in Figure 10-9. Note that this linking process appears somewhat complex, even though only a few linkages are depicted. In a real system, the student file will likely contain several thousand students and the class schedule file will likely contain several hundred classes. If the university decides to integrate a student fee file, faculty file, and facilities file (i.e., buildings and rooms) with the student and class schedule files, the linkages between files will become even more complex. Herein lies the problem with tra-

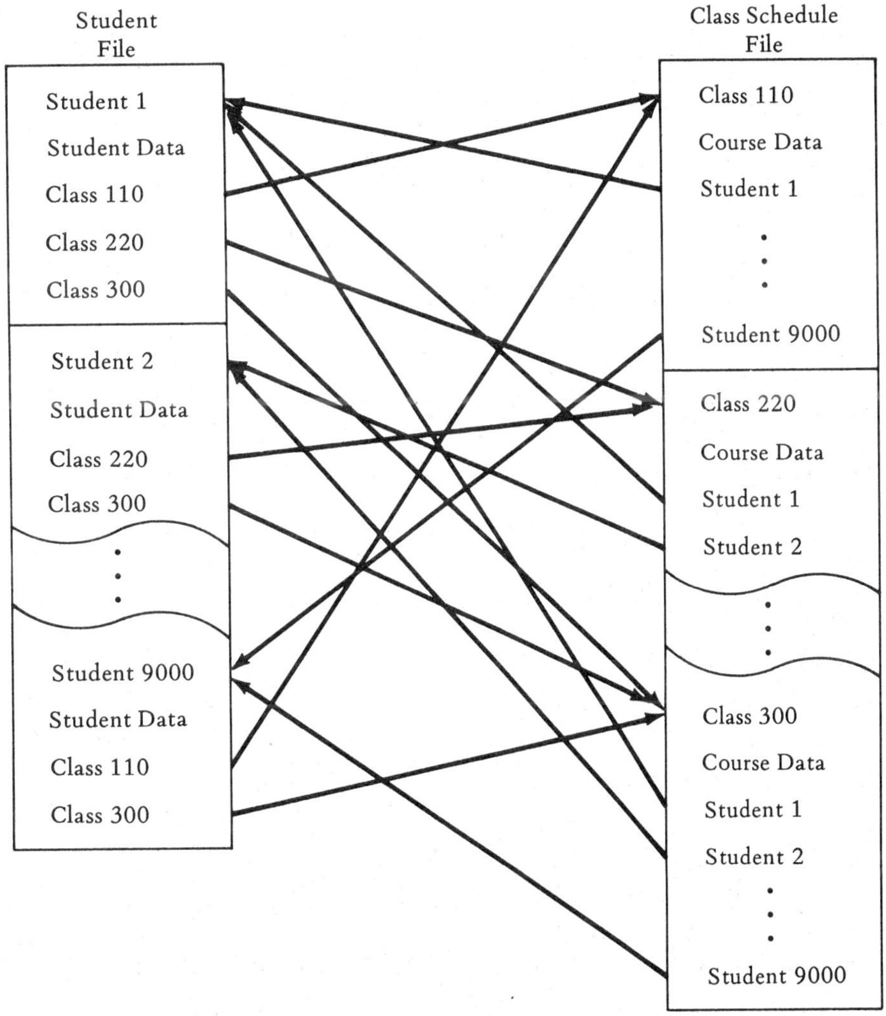

Figure 10-9. *Linking Data between a Student File and a Class Schedule File*

ditional approaches to integrating data from separate files. First, programming such complex relationships is a massive task. Second, since the needs for all relationships and data elements are to a great extent unforeseen and/or continually changing, additional data elements and relationships must be added and programmed into the system. A simple adding of a few data elements to a file can necessitate modification of several dozen programs that access the file. Even when data integration is achieved through file consolidation, system modifications can have serious rippling effects throughout computer programs involved in the system. Consequently, many organizations have found up to 80 percent of their programming staff's efforts are directed toward making modifications to existing systems rather than toward developing new ones.

DATABASE TECHNOLOGY

The first section of this chapter defines and describes the major factors contributing to the difficulties of integrating organizational data for management information. These factors are reviewed below.

1. Some data are not suited for integration or do not need to be integrated.
2. Organizations often fail to consider data integration when designing and developing their information systems.
3. There is considerable technological complexity in integrating organizational data through consolidation of files and/or linking data in separate files.

The first factor is one that simply needs to be recognized. Organizations must accept the fact that not all organizational data can be neatly integrated into a single, all-encompassing database. The second factor can be overcome by better planning and more top management involvement in the design and development of organizational information systems. The third factor reflects technological constraints. An organization that has overcome the first two factors generally knows the right way to approach the development of integrated information systems. However, the organization may need help in doing it. A database management system (DBMS) provides the technology to more effectively and efficiently achieve integrated information systems.

Database

When organizational data are interrelated and stored together to serve one or more applications, the collection of data is generally referred to as a database. Data redundancy is seldom totally eliminated, but it is controlled and significantly reduced. Figure 10-10 is a simplified view of files integrated into a database structure. An organization can have more than one database structure. For example, a university may have a database for a student information system consisting of student, course, and facilities files and a database for a financial information system consisting of general ledger, accounts receivable, accounts payable, and payroll files. When an organization decides to have two or more databases, the databases are disjoint in structure.

As discussed earlier in the chapter, programming data interrelationships to achieve integration and reduce redundancy is a complex task. Computer programs must use one or more of the file access techniques described in Chapter 4 (i.e., sequential, random, or indexed) to provide explicit instructions as to the physical locations of the data elements required to satisfy a particular online terminal query or to produce a particular batch report. Figure 10-11 shows the relationship among application programs, file access software, and computer files.

For example, a program may need to access several interrelated data elements from records located in different files. The program must use indexes to compute physical addresses for each

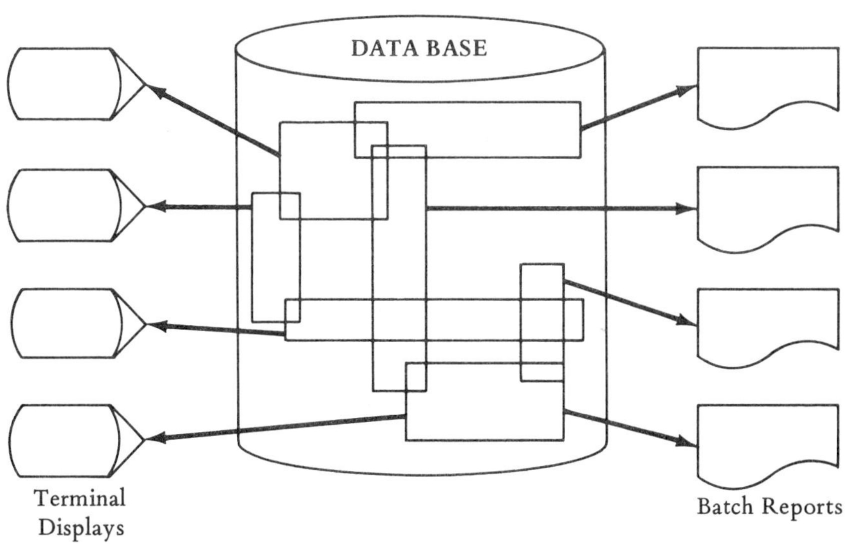

Figure 10-10. *Simplified View of an Integrated Database Using Batch and Online Processing*

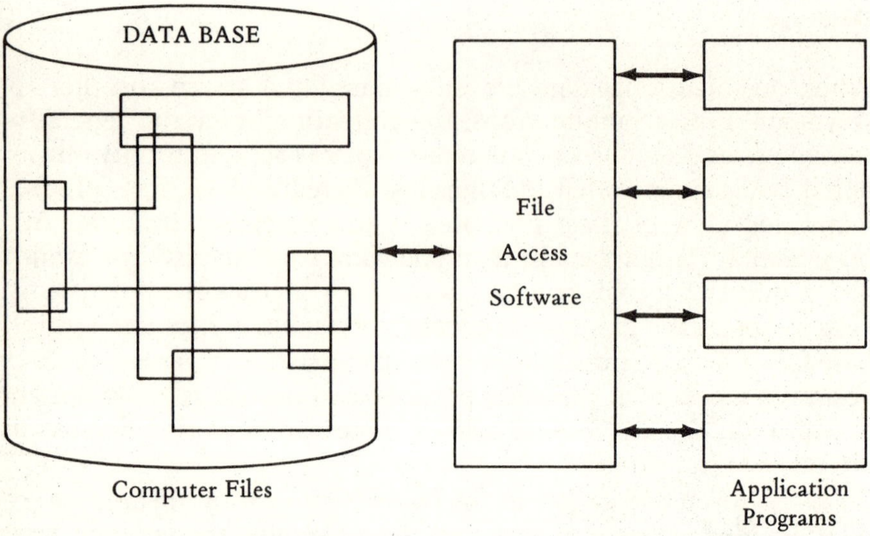

Figure 10-11. *The Relationship among Application Programs, File Access Software, and Computer Files*

file in which data are contained, retrieve the entire record from each file, and then extract the data elements required from each record. Accordingly, programs must be coded to consider the physical locations and structures of all files, records, and data elements in the data base. If a new data element is added to a record, resulting in a new record format, all programs accessing that record must be modified (even those to which the new data element is irrelevant). Ideally, application programs should be able to access the data in a data base, independent of the physical structure of the data in the data base. In other words, a program should be able to ask for and get only the data elements needed without having to get and process the entire records in which the data elements are located.

DBMS

During the early 1970s, generalized software packages were developed by various companies to reduce the difficulties of integrating the data contained in a database. These software packages are referred to as database management systems (DBMS's). Their developers have made a significant contribution to the design and implementation of integrated computer-based information systems.

A DBMS can be envisioned as an additional layer of software between file access software and application programs (see Figure 10-12). It is beyond the scope of this text to explain the actual

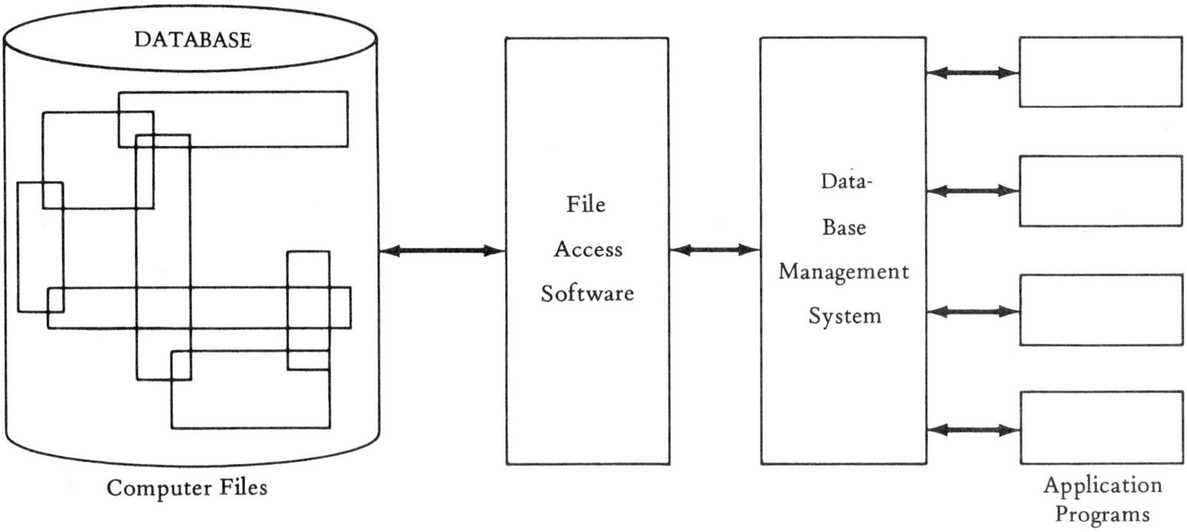

Figure 10-12. *The Role of a DBMS*

approaches and resulting mechanics of the DBMS's available.[3] Their functions are described below to the extent needed for a systems analysis perspective.

A DBMS removes from the application program the function of explicitly coping with the physical locations and structures of the files and records in which data elements are stored. Therefore, the application program need only indicate the specific data elements required along with the key(s) or index(es) necessary to uniquely identify the group of data elements. With a DBMS, any data element or combination of data elements can be used to index a group of data from a record or records. (This is, usually accomplished using an automatic inverted indexing scheme, as discussed in Chapter 4.) The DBMS, in conjunction with the file access software, handles the necessary logistics of retrieving and storing specified data elements in the proper physical locations within the database.

Many vendors and software companies offer sophisticated advancements to their DBMS software packages in the form of query languages. A query language is a set of easy-to-use computer instructions. These instructions are designed to allow a person who is not trained in computer programming to retrieve, modify, add, or delete selected data elements, based upon stated conditions.

3. Readers interested in an excellent treatment of database management systems should see one of the following texts: *An Introduction to Database Systems*, by C. J. Date, 1975; *Principles of Data-Base Management*, by James Martin, 1976; and *Database Processing*, by David Kroenke, 1977.

The use of query languages is becoming increasingly popular due to the proliferation of computer terminals, ad hoc retrieval requests from all levels within an organization, and the ever-improving software packages readily available for most user applications. Such advancements place the computer in the hands of the user and help remove the aura of mystery that surrounds computers.

Thus, a DBMS allows the users of a database to view the database in terms of the application necessary to support their departmental or functional area. Their access to the database is not complicated by other data elements in the database. This means that data elements can be added to or deleted from the database, yet affect only the application programs that use those data elements. A DBMS greatly reduces the difficulty of both establishing and modifying integrated files.

SYSTEMS ANALYSIS IN A DATABASE ENVIRONMENT

A database approach to information systems development does have effects on systems analysis and systems administration. To achieve integration and the control and reduction of data redundancies, much more coordination and control of the systems development process is required.

Systems Administration

Most organizations are in the fourth stage of EDP growth before they attempt to integrate organizational data into one or more databases (review Table 9-1). During the preceding third stage of EDP growth, most organizations establish an MIS steering committee composed of top-level managers. This top management involvement and perspective is crucial to insuring a systematic view and coordination of information systems activities in a database environment. Since top management represents the point where the organization is physically integrated, the managers at this level can best direct the proper integration of organizational data.

Database Administration

The integration of data into one or more centralized databases requires centralized coordination and control of data elements and their structure. For example, since data elements are referred to by data names, care must be taken to insure that each data element within a database is assigned a unique name.

To help insure proper administration of a database, the position of database administrator has been established in many organizations. The database administrator is primarily involved in planning and controlling the overall structure and contents of an organization's database(s). (For review, see "Personnel" and Tables 4-1 through 4-3 in Chapter 4.)

The overall structure and contents of a database are referred to as its *schema.* A more limited view of the database taken by an individual application is referred to as a *subschema.* That is, each application uses only a portion of the overall database. The database administrator must plan and control the schema as well as the various subschemas of the database. The integrity and proper operation of the overall database and of the multiplicity of computer programs that access subsets of the database are concerns of the database administrator.

Data Element Dictionary

The contents of the database are defined by a data element dictionary. This dictionary is similar to the one described in Chapter 6. It documents each data element's name, definition, source, where used, maintenance, and storage. Since a database generally serves several application systems, there is a need to restrict access to data elements. The database dictionary is usually expanded to define who is authorized to access what data elements.

Selecting a DBMS

The selection of a DBMS is a major, often difficult, decision. As indicated earlier, a number of DBMS packages have been developed by computer vendors and independent software companies. The more popular DBMS packages and their suppliers are listed below:

- TOTAL/7
  Cincom Systems
- Information Management
  System (IMS)
  IBM
- IDS/II
  Honeywell Information
  Systems

- DMS II
  Burroughs
- MARS VI
  Control Data Corporation
- SYSTEM 2000
  MRI Systems
- ADABAS
  Software AG
- DMS 1100
  Sperry Univac

These DBMS packages have a variety of capabilities and resource requirements. The selection of a particular DBMS should be based on organizational requirements and the cost effectiveness of the alternative DBMS packages available. The selection process should be pursued as defined in Chapter 7.

Systems Analysis

The use of a DBMS significantly enhances the effectiveness and efficiency of the information systems developed by systems analysts. The DBMS makes it easier to integrate the collection, processing, storing, and reporting of organizational data. Another benefit of a DBMS is that it generally leads to increased involvement by top management.

The process of systems analysis is not significantly affected by the implementation of a DBMS. As described in Chapter 6, the systems analyst continues to develop systems specifications—output definitions, input definitions, decision tables, and flowcharts. However, in a database environment, the data element dictionary is controlled by the database administrator(s). Accordingly, as a systems analyst determines the data elements necessary to support the system, he or she reviews them with the database administrator. Data elements that already exist in the database can be made accessible to the information system. This data sharing is one of the primary means of controlling data redundancies. Any new data elements required for the information system can be assigned unique names and entered into the database under the supervision of the database administrator. After all new data elements have been defined, the database administrator can handle any physical structuring required to incorporate the new data elements in the database.

SUMMARY

One of the early expectations of organizations establishing computer-based information systems was that the systems would be able to integrate all organizational data into a single, all-encompassing database to serve management's information requirements. However, the integration of organizational data has been constrained by the failure to plan for integration during systems development, the fact that some organizational data are not suited for integration, and the technological difficulties of integrating data. When organizational data are not integrated, redundancy in data collection, processing, storage, and reporting occur.

The integration of data requires that data files be consolidated and/or logically linked together. The programming difficulties of this integration have been greatly reduced by the advent of DBMS's. A DBMS handles the logistics associated with the physical structure of the data. The application programmer can concentrate on the necessary logical structures.

To help insure proper planning and control of an organization's database(s), a database administration function is often established. The database administrator makes certain that all naming, defining, linking, and accessing of data elements are conducted in a centralized orderly manner. In a DBMS environment, the systems analyst coordinates his or her need and use of data elements in the context of the overall database.

EXERCISES

1. Read "MIS Is a Mirage," *Harvard Business Review,* Jan–Feb 1972, pp. 90–99, and discuss the strengths and weaknesses of John Dearden's position.
2. What characteristics of higher-level-management decision making have made it difficult to provide higher-level management with computer-based information systems support?
3. How does redundancy occur in an organization's data files, and what problems does it cause?
4. Discuss the importance of integration of organizational data.
5. Discuss the difference between the consolidation and linking methods of achieving integration of data.

6. Discuss the relationship between a DBMS and file access techniques.
7. What are the organizational and operational implications of doing systems analysis in a database environment?

SELECTED REFERENCES

Canning, R. G. "Trends in Data Management," Part 1, *EDP Analyzer* Vol. 9, No. 5 (May 1971); Part 2, *EDP Analyzer* Vol. 9, No. 6 (June 1971).

CODASYL Systems Committee. *A Survey of Generalized Data Base Management Systems*. Technical Report. May 1969.

CODASYL Systems Committee. *Feature Analysis of Generalized Data Base Management Systems*. Technical Report. May 1971.

Date, C. J. *An Introduction to Database Systems.* Reading, Mass.: Addison-Wesley, 1975.

Dearden, John. "MIS Is a Mirage," *Harvard Business Review*, Jan–Feb 1972, pp. 90–99.

Engles, R. W. "A Tutorial on Data Base Organization," *Annual Review in Automatic Programming*. Vol. 7, Part 1. ed. Halpern and McGee. Elmsford, N. Y.: Pergamon Press, July 1972, pp. 1–63.

Everest, G. C. "The Objectives of Database Management," *Proc. 4th International Symposium on Computers and Information Science.* New York: Plenum Publishing Corporation (in press).

Kroenke, David. *Database Processing.* Chicago: Science Research Associates, 1977.

Martin, James. *Principles of Data-Base Management.* Englewood Cliffs, N. J.: Prentice-Hall, 1976.

Meltzer, H. S. "Data Base Concepts and an Architecture for a Data Base System," Presentation given to SHARE Information Systems Research Project, SHARE XXXIII, Boston, Mass., Aug 20, 1969.

Nolan, Richard L. "Computer Data Bases: The Future Is Now," *Harvard Business Review*, Sept–Oct 1973, pp. 98–114.

Schussel, George. "The Role of the Data Dictionary," *Datamation*, June 1977, pp. 129–148.

11 Management Information Systems and Decision-Support Systems

In Chapter 10, one of the reasons cited for management's disappointment in the concept of MIS was the limited support that computer-based information systems have provided for top-management decision making. The fact that decision-support systems (DSS) are increasingly providing capabilities that are significantly offsetting this disappointment was also pointed out. These capabilities represent an advanced dimension of computer-based information systems. The systems are more useful and better accepted by operating, middle, and top management. A DSS goes beyond transaction processing and reporting. It actually assists management in making decisions. This dimension of information systems is at a formative (but, certainly, a progressing) state.

Herbert A. Simon's model for describing the decision-making process was introduced in Chapter 5.[1] This model is shown in Figure 11-1. Decision making is a three-phase, continuous process, flowing from intelligence to design to choice but capable of returning to a previous phase when necessary. The need for a decision-making process is invoked by the recognition of a problem or an opportunity. The resultant decision-making activity is directed at resolving the problem or taking advantage of the opportunity.

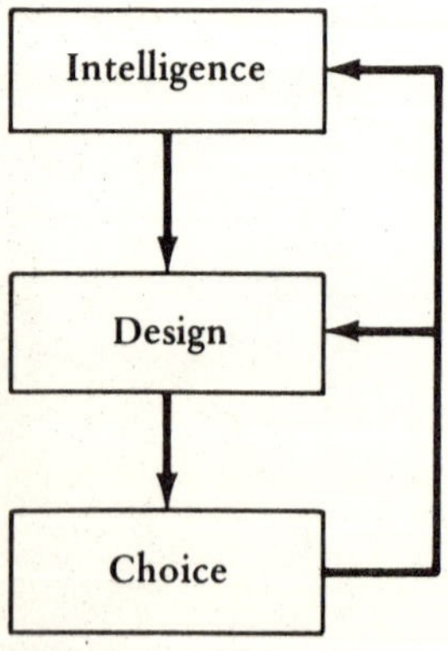

Intelligence

Recognizing problems or opportunities calling for decisions. Gathering and structuring information for enlightenment about the problem or opportunity.

Design

Developing and analyzing alternative solutions.

Choice

Selecting a particular solution from among those available for implementation.

Figure 11-1. *The Decision-Making Process*

1. Herbert A. Simon, *The New Science of Management Decision,* 1960, p. 54.

A DSS functions to directly support the intelligence and design phases of decision making. In so doing, a DSS indirectly supports the choice phase. However, a DSS does not actually make the choice. When a system makes choices for management, it moves out of the realm of a DSS and into the realm of programmed decision making.

The topic of DSS is discussed in four sections of this chapter. The first section reviews management decision making and its relationship to DSS. The second and third sections discuss DSS as they pertain to the intelligence and design decision-making phases, respectively. The fourth section discusses the impact of DSS on choice.

It is not the objective of this chapter, nor would it be realistic, to provide preparation for designing and developing a DSS. Rather, the objective is to provide exposure to the nature and potential of a DSS. Hopefully, the chapter will stimulate interest in further study of techniques required to develop a DSS. A systems analyst who contributes in the area of DSS must be skilled in such areas as management science (synonymous with decision science and operations research), modeling, simulation, and advanced statistics. To comfortably read the material in this chapter, one needs only a basic background in statistics and mathematics. The selected references at the end of the chapter are sources of more advanced discussions of the topics presented.

DECISION MAKING

The nature and characteristics of decision making are concepts discussed frequently in this book. In this section of the chapter, these concepts are reviewed and expanded to provide a framework for discussing DSS.

Characteristics of Organizations and Organizational Processes

In Chapter 3, it was pointed out that organizations and organizational processes can be characterized as existing on a continuum between closed/stable/mechanistic and open/adaptive/organic. The more closed/stable/mechanistic an organizational process is, the more structured, computational, and routine is the decision making associated with planning and controlling it. Consequently, the

Table 11-1. *Characteristics of Decision Making and Information at Different Management Levels*

	Management Level		
	Operating	*Middle*	*Top*
Characteristics of Decision Making	Computational/Objective ——————→		Judgmental/Subjective
Examples	Inventory Reordering Production Scheduling Credit Approval	Short-Term Forecasting Budget Preparation Capacity Planning	New Product Planning Location of New Factory Mergers and Acquisitions
Characteristics of Information	Internal/Complete ——————→		External/Incomplete
Examples	Sales Order Production Requirements Customer Credit Status	Sales Analysis Budget Analysis Production Summaries	Market Conditions Industry Forecasts Government Regulations

decision making associated with closed/stable/mechanistic processes can be more readily supported with computer processing.

Generally, higher levels of management are dealing with more open/adaptive/organic decisions, increasing the difficulty of making a good (let alone, a perfect) decision. The lack of structure and routine, as well as the external and incomplete nature of information, at higher organizational levels of decision making significantly constrains the ease of providing computer support. Table 11-1 (duplicated from Chapter 3) reviews both the characteristics of decision making and the information required at different organizational levels.

Knowledge of Outcomes

An important dimension of decision making is the degree of certainty with which decision makers can view the outcomes or results of their decisions. The knowledge of the outcomes of decision making can be categorized as follows:

1. **Decision making under certainty.** There is complete, accurate knowledge of the outcome of each choice available. Each decision has only one unique outcome (or set of events) associated with it.

2. **Decision making under risk.** Alternative decisions can result in more than one outcome. The possible outcomes and their probabilities of occurrence can be identified.
3. **Decision making under uncertainty.** Alternative decisions can result in more than one outcome. Some of these outcomes can be identified but their probabilities of occurrence are unknown.

There is a relationship among the degree of certainty in decision making, the characteristics of organizational processes, and management levels. This relationship is depicted in Figure 11-2. The relatively open/adaptive/organic processes that higher-level management must cope with increase the uncertainty associated with decision making. Conversely, the relatively closed/stable/mechanistic processes that lower-level management must cope with reduce the uncertainty associated with decision making. The examples in Table 11-1 convey the varying degrees of uncertainty associated with decision making at different management levels.

In decision making under certainty, the objective is to determine which is the best solution to a problem or the best opportunity. Since the decision variables and their relationships are known, the best solution can usually be computed. Accordingly, tools based upon mathematics are useful. In decision making under risk, the problem is to determine, within acceptable levels of probability, the decision that provides an optimal (or, more likely, a satisfactory) outcome. Accordingly, tools based on probability or statistics are useful for decision making under risk. In decision making under uncertainty, the probabilities of outcomes are not

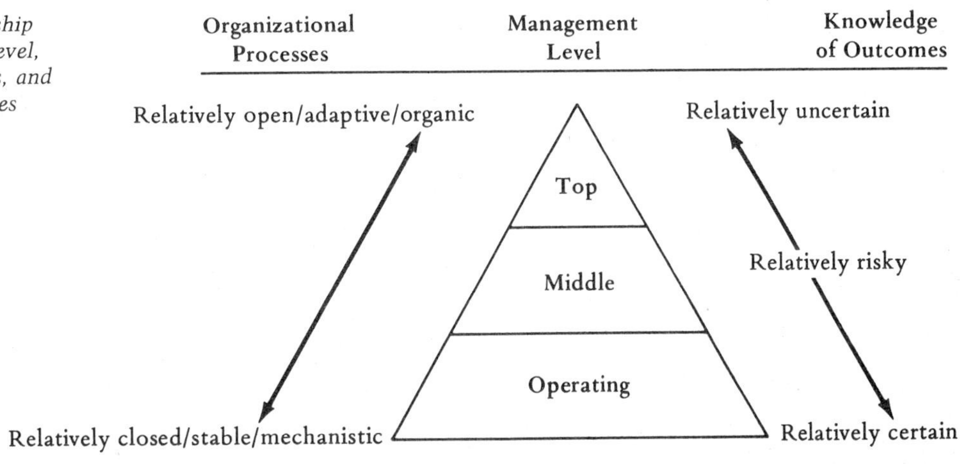

Figure 11-2. *Relationship among Management Level, Organization Processes, and Knowledge of Outcomes*

known. Therefore, the decision maker must assign arbitrary probabilities or assume no probabilities. In the former case, statistics can be used. In the latter case, decision making is not based on quantitative techniques.

A more extensive discussion of the use of mathematical, statistical, and other decision-support tools is presented later in the chapter.

Behavioral Considerations

The cognitive style of the decision maker is an often overlooked but significant dimension of the decision-making process. The cognitive style of decision makers can be expressed as existing on a continuum ranging from intuitive/heuristic to systematic/analytic.[2] The key characteristics of these styles are summarized below.

Intuitive/Heuristic

Intuitive/heuristic-oriented managers tend to learn from experience. They use common sense and intuition to approach decision making in a trial-and-error mode. They tend to view a problem or an opportunity as a totality rather than as a structure constructed from specific parts. Highly visible situational differences that vary with time are the basis from which they draw inferences about different situations.

Systematic/Analytic

Systematic/analytic-oriented managers tend to learn by analyzing a problem or an opportunity. They do not rely heavily on previous experiences and feedback. Rather, decision making is based on formal rational analyses in which explicit, quantitative models are often used. These managers prefer to reduce a problem or an opportunity to a set of underlying causal functions and then look for similarities or patterns by analyzing and comparing variables.

---

2. These terms and concepts are based upon combining the results of research findings from the following authors. James L. McKenney, *Humans as Information Processing Systems.* Working Paper 72-4. Harvard Graduate School of Business Administration, Boston, 1972. Michael Barrett, *Information Processing Types and Simulated Production Decision Making.* Working Paper 73-2. The MIS Research Center, University of Minnesota, Minneapolis, 1973.

Implications for DSS

The cognitive styles of managers affect the potential contribution a DSS can have on their decision making. Systems analysts, out of necessity, tend to be systematic/analytic. The highly structured and logical computer technology they work with demands that particular style. Consequently, systems analysts often perceive all decision making as systematic/analytic. However, systems analysts must recognize and be sensitive to the cognitive styles of the managers for which a DSS is designed.

A DSS that includes highly structured, predetermined sequences of analytical tools will likely be well received by systematic/analytic managers but may not be used by intuitive/heuristic managers. Rather, the latter may only be interested in a DSS that allows them to explore a wide set of alternatives that depart from predefined models. Further, intuitive/heuristic managers may not be at all interested in computer assistance in their decision-making process. They often prefer to rely on their hunches and/or information based on discussions with peers or subordinates. In such cases, systems analysts should accept that managers may not desire or use a DSS, irrespective of its merit.

When possible, and when management is willing, a systematic/analytic approach is usually preferable to an intuitive/heuristic approach. However, there are many cases where an intuitive/heuristic approach is appropriate, or even necessary, irrespective of management's cognitive style. For extremely open/adaptive/organic problems or opportunities, when there is limited time, or when a systematic/analytic approach is too expensive relative to the importance of the decision, an intuitive/heuristic approach is usually preferable.

For example, consider a company that has an option to buy the patent rights for a revolutionary solar energy device. The company is not certain of the marketability of the device or if other superior devices are forthcoming. The patent option costs $100,000. Potentially, the company could make several million dollars. To adequately research and prepare for this decision requires three months of work at a cost of $120,000. Nevertheless, the company must make a decision within two days or lose its option to a competitor.

To summarize, systems analysts should strive to consider cognitive styles of managers and design DSS's that are flexible enough to accommodate different managerial approaches. They should also accept that, due to the limitation of DSS technology and/or the limitations of managers, many good and poor decisions are made without the potential assistance available through computer technology.

DSS AND DECISION INTELLIGENCE

The intelligence phase of decision making involves becoming aware of problems or opportunities. Initially, this awareness is usually at a symptomatic level. Therefore, the intelligence phase usually requires delving deeper into the problem or opportunity to fully identify and examine the underlying variables and their relationships. For example, a symptom of a problem may be a decrease in sales, but the actual cause of the problem may be a change in consumer behavior, a new product introduced by competitors, a morale problem in the sales force, or a combination of these factors or additional ones.

The more intelligence that management has of the cause of a problem or the effort required to capitalize on an opportunity, the more success it is likely to have in designing and choosing an optimum, or at least a good, decision. Indeed, it is the degree of intelligence that to a great extent determines whether a particular decision will be made under certainty, risk, or uncertainty.

Exception and Summary Reporting

Since the first step of decision making is identifying problems or opportunities, it logically follows that a DSS should assist in this identification process. The routine reporting provided by an information system plays a key role in providing management with intelligence about problems or opportunities. A common and successful means of identifying a great deal of organizational problems and opportunities is the use of exception reporting derived from routine transaction processing of the organization's internal database. For example, inventory items that experience an unanticipated decrease or increase in activity may be identified as problems or opportunities. Another means for identifying problems and opportunities is through the use of summarized information. In particular, graphical presentations of summarized data have gained great popularity. Some board of director conference rooms are now equipped with video equipment that graphically displays various cost and profit trends from data generated from computer systems. Unusual shifts in summarized data (either favorable or unfavorable) alert management to further investigate detail data to assess the causes of these shifts.

Both exception and summarized reporting have the advantage of presenting information in a decision-impelling format. That is, they highlight important information that encourages management action rather than overload management with details.

DBMS

A constraint of exception, summarized, and detail reports is that their content and format must be determined in advance (i.e., so they can be programmed). However, the open/adaptive/organic characteristic of many organizational processes makes it extremely difficult to anticipate all information requirements. Consequently, managers frequently request crash projects to produce special reports for decision making. Unfortunately, the complexity and frequency of such requests often make it impossible for information systems personnel to respond to the requests within the time constraints placed upon them.

With the advent of DBMS's, the ability to query a data base in ad hoc (rather than predetermined) ways has been greatly enhanced. Most DBMS's include a general-purpose query language that can be used for information retrieval purposes by persons who are not technicians. These languages allow users to specify the values or ranges of attributes to be used for including and/or excluding a group of data for a particular query. For example, a credit manager may easily specify that he or she needs a list or display of all credit customers as follows:

```
Include if:
    BALANCE-DUE > 500
    DAYS-PAST-DUE = 30 to 90
    SALES-DISTRICT = 4
Exclude if:
    VIP-CODE = 1
```

The query language facility is an additional layer of software between the user and the DBMS. This relationship is shown in Figure 11-3. Because of the general-purpose nature of query languages, queries tend to be less efficient in execution than application programs. Also, query languages generally have limited report formatting capabilities. Therefore, frequently required reports

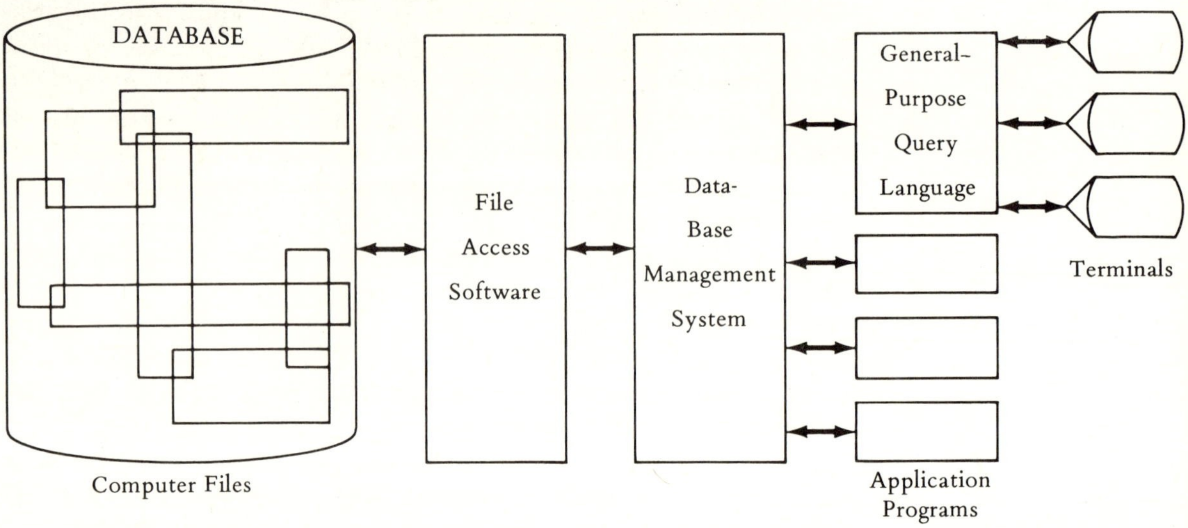

Figure 11-3. *Relationship of Query Language to Database Management System*

should continue to be programmed using a procedural language (e.g., COBOL or PL/I). The query language should be used for reporting that is ad hoc and infrequent.

Statistics

During the intelligence phase of decision making, management attempts to reduce, to the extent possible, the uncertainty associated with decision making. Statistical analysis is a useful tool for developing a better understanding of variables surrounding a decision-making process. Statistics can be used to structure unorganized detail data into descriptive measures such as range, mean, mode, median, and frequency distributions. Analyses can be conducted on such data as employee salaries, total sales, production yields, and the like. They can be useful for analyzing a group of data or for comparing one or more groups of data (e.g., comparing sales in different stores).

Management, particularly top management, does not usually have access to all data pertinent to a decision. Consequently, management must frequently deal with incomplete information. This is especially true when the information needed is external to the organization. For example, it is not practical to survey all potential consumers of a new economy car to determine their attitudes toward its styling, engineering, and performance. However, management can take a statistical sample of the entire population of potential consumers and use inferential statistics to generalize about overall consumer behavior.

In other cases, management may want to determine the similarities and differences between two or more populations for which data on only a sample of each population can be collected. Again, statistical tests can be used to compare the samples and generalizations can be inferred about the similarities and differences. Such tests include *t*-tests, chi-square, and analysis of variance.

A number of statistical packages have been developed. They provide an excellent repertoire of commonly used statistical tools. The packages can be used to conduct complex statistical analyses with minimal effort. The addition of one or more statistical packages to an organization's computer-based information systems can greatly enhance the ability to make prompt statistical analyses in support of decision making. Four of the commonly used statistical packages and their suppliers are listed below.

- BMD (Biomedical Computer Programs)
  UCLA
  Los Angeles, California
- Omnitab
  National Bureau of Standards
- SAS (Statistical Analysis System)
  SAS Institute Incorporated
  Raleigh, North Carolina
- SPSS (Statistical Package for the Social Sciences)
  SPSS Inc.
  Chicago, Illinois

In practice, few managers use statistical packages directly. Rather, their staff and/or members of the information systems staff perform the necessary technical tasks.

DSS AND DECISION DESIGN

The design phase of decision making involves developing and analyzing decision alternatives. In recent years, numerical models have gained in popularity as aids for decision making, particularly among managers with systematic/analytic orientations. Numerical models are quantitative representations of objects or processes. They can be visualized as systems capable of transforming input data about an object or process into informative output about an object or process. The transformation process is conducted in accordance with one or more processing statements (e.g., formulas) defined within the model.

Consider as a simple numerical model the mathematical equation used to compute the area of a rectangle.

$$\text{Area} = \text{Length} \times \text{Width}$$

The imputs to the model are values for the variables length and width. The transformation process consists of multiplying length times width and assigning the resulting value to the variable area. The output from the model is the computed area for a given length and width. This model can be used to compute the area of any rectangle for which the width and length are known. This general-purpose characteristic of models makes them extremely useful. Once a model has been developed to represent an object or process, it can be used repeatedly for similar applications, no matter how complex the model is.

The use of mathematical equations provides considerable precision and convenience. For example, mathematical equations can be used to compute the areas of triangles, circles, squares, and so on. However, many objects or processes are difficult or impossible to represent using mathematical equations. Consider the irregularly shaped object in Figure 11-4*a*. There is no convenient or accurate mathematical equation available to compute the area of this or similarly structured objects. However, the area can be reasonably estimated by using numerical techniques of probability as described below:

1. Enclose the irregularly shaped object in a square.
2. Place the numbers 1 through 10 at equally spaced intervals on the *Y*- and *X*-axes of the square.
3. Place ten pieces of paper numbered 1 through 10 in a hat.
4. Randomly generate pairs of *X* and *Y* values by repeatedly drawing numbers from the hat, being careful to return each number to the hat and to shuffle all numbers before the next drawing. This insures that each number has an equal chance of being drawn each time.
5. Plot each pair of *X* and *Y* values using the scales of their respective *X*- and *Y*- axes. For example, the following values for *X* and *Y* are plotted in Figure 11-4.

Point	*A*	*B*	*C*	*D*	*E*
X	2	4	7	9	8
Y	6	4	8	5	4

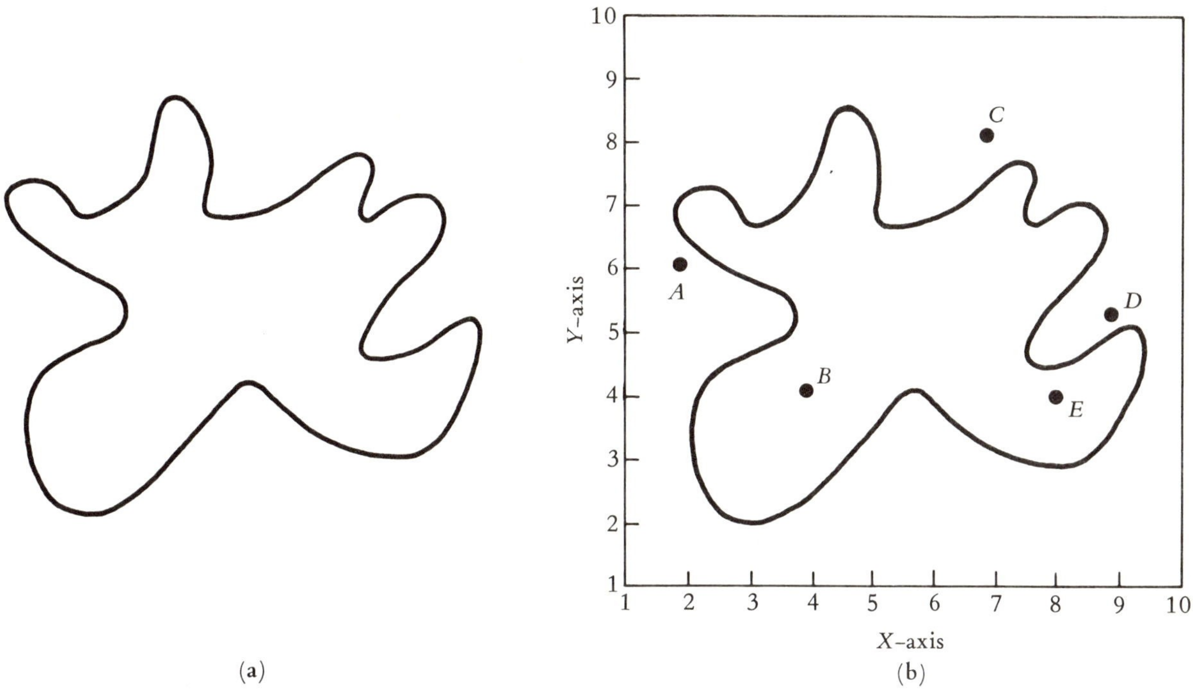

Figure 11-4. *Using Probabilities to Estimate the Area of (a) an Irregularly Shaped Object with the Area of (b) Randomly Selected Coordinates*

6. After several (X, Y) coordinates have been plotted, determine the percentage of points located within the irregularly shaped object.
7. Compute the area of the square and multiply it by the percentage of points within the irregularly shaped object.
8. The result is an estimation of the area of the irregularly shaped object.

If this entire process is repeated several times, the computed area estimates are likely to differ slightly. This variability in outputs is due to the use of probabilities in the modeling process.

Numerical models based strictly on equations always generate the same output, given the same input. They are called *deterministic models.* Numerical models that contain dimensions of probability produce different, though usually predictable, outputs. These models are called *stochastic models.*

Both deterministic and stochastic models are used to support management decision making. Certainly, these models are of greater complexity and size than the preceding two examples. Fortunately, the computational capabilities of computer technology allow complex, large models to be manipulated conveniently.

The execution of a model to represent an object or process is called *simulation.* Because decision making is concerned with outcomes, the ability to simulate outcomes of possible situations is particularly helpful in designing decision alternatives.

The ability to simulate the results of decisions using a model is a powerful tool. Models are usually much less expensive to experiment with than are the objects or processes they represent. Therefore, decisions are tested and outcomes analyzed using the model rather than the real objects or processes.

For example, the management of a hotel is planning to re-carpet the entire hotel. Management needs to know how much carpet is required and how much it will cost. It would be foolish to buy the carpet and then see how much carpet is required. Rather, hotel employees can measure the lengths and widths of all rooms and then use the model for computing area to determine the total number of square feet of carpet required. By including the cost per square foot of carpet as an input variable to the model, they can simulate what it will cost to carpet the entire hotel using various grades of carpet. The equations within this deterministic model are:

$$\text{Area} = \text{Length} \times \text{Width}$$

$$\text{Total Cost} = \text{Area} \times \text{Cost per Square Foot}$$

The type of model constructed and the simulation conducted for a particular decision are a function of whether the decision is made under certainty, risk, or uncertainty. Modeling and simulation for these decision-making categories are discussed below.

Models for Decision Design under Certainty

Decision making under certainty pertains to situations where there is only one outcome for each decision alternative. Therefore, models used to simulate outcomes for decisions made under certainty must produce only one outcome for each decision alternative. Accordingly, deterministic models based on mathematical equations are used.

Usually, the objective of decision making under certainty is to optimize some objective function (e.g., to maximize productivity or minimize costs). Several general-purpose mathematical models for computing optimum solutions for a variety of organizational functions are available. To illustrate their type and variety, some of the models are listed below; however, it is beyond the scope of this chapter to explain these techniques.

Input	Model	Output
Sales Operating Expenses	Cost of sales = 0.35 X sales Gross margin = sales − cost of sales Profit before taxes = gross margin − operating expenses Taxes = 0.48 X profit before taxes Net profit = profit before taxes − taxes	Net Profit

Figure 11-5. *FI Model for Computing Net Profit*

1. Systems of equations
2. Linear programming
3. Integer programming
4. Dynamic programming
5. Break-even analysis
6. Return on investment
7. Queueing models
8. Inventory models

In many cases, models must be tailored to particular organizational processes. Consider the simplified model for computing net profit in Figure 11-5. The model requires two inputs: sales and operating expenses. A manager can use a terminal to interact with a computer-program representation of the model. He or she enters either actual or predicted values for the two variables and receives simulated profit statements. Figure 11-6 shows how this person/machine dialogue progresses. By simply re-executing the model, the manager can assess the effects of different levels of sales and operating expenses on net profit.

Manager:	EXECUTIVE PROFIT-MODEL	
Terminal:	ENTER SALES	
Manager:	125,750	
Terminal:	ENTER OPERATING EXPENSES	
Manager:	58,100	
Terminal:	PROFIT STATEMENT	
	SALES	$125,750
	LESS: COST OF GOODS SOLD	44,013
	GROSS PROFIT	$ 81,737
	LESS: OPERATING EXPENSES	58,100
	PROFIT BEFORE TAXES	$ 23,637
	LESS: TAXES	11,346
	NET PROFIT	$ 12,291

Figure 11-6. *Use of a Computer-Supported Model for Computing Net Profit*

Input

Sales

Long-Term Debt
Average Short-Term Debt

Accounts Receivable
Balance at Beginning
of Period

Model

Administrative expense	= 0.08 × sales
Advertising expense	= 0.04 × sales
Interest expense	= 0.08 × long-term debt + 0.09 × average short-term debt
Bad-debt expense	= 0.01 × accounts receivable balance at beginning of period
Operating expenses	= Administrative + advertising + interest + bad-debt

Output

Operating Expenses

Figure 11-7. *A Model for Computing Operating Expenses*

The design and development of DSS models usually begins with simple models calling for high-level inputs. As a better understanding of a process emerges, basic models are expanded and/or modules are added to provide greater detail and capability. The model can then generate high-level items from more basic inputs.

For example, operating expenses have to be inputted to the model in Figure 11-6. However, an additional module or another model could be developed to compute operating expenses. Figure 11-7 shows such a model. The additional inputs (long-term debt, average short-term debt, and accounts receivable balance) could be generated by another model, and so on. By interfacing such model, a sophisticated profit model can be established.

Other models can be developed for other organizational processes—financing, staffing requirements, facility requirements, and so on. This is accomplished by developing mathematical equations that are representative of the processes and then translating them into computer programs.

Models for Decision Design under Risk

Decision making under risk pertains to situations where alternative decisions can result in more than one outcome and the probabilities of these outcomes are known. This means that at least one dimension of the object or process under consideration has an element of chance associated with it. Therefore, models used to simulate outcomes for decision making under risk must be able to represent elements of probability within the model to reflect the probabilities associated with outcomes. Accordingly, stochastic models using numerical techniques based on probability are used for decision making under risk.

Stochastic models generally contain mathematical equations along with probability functions. To the extent that a process can be represented mathematically, it is advantageous to do so because of the greater precision afforded. Probability functions should be used only when a process cannot be conveniently and accurately represented by one or more mathematical equations.

Considerable perception is required to develop a mathematical equation or set of equations to model an object or process. In many cases, it is not possible, given existing mathematical techniques, to model objects or processes. In other cases, objects and processes are modeled using probability functions until they can be described mathematically. In such cases, the probability functions serve as a surrogate until the mathematical functions can be defined.

As a fictitious illustration of this sequence, assume that the mathematical equation for computing the area of a triangle is unknown. The probability technique illustrated in Figure 11-4*b* can be used to estimate the area of a triangle. Once it is discovered that a mathematical equation (area = ½ × base × height) can be used to compute the exact area of a triangle, it is no longer necessary, and certainly not advantageous, to continue using the probability technique.

Many organizational processes are difficult or impossible to describe by mathematical equations. Therefore, techniques of probability are the norm rather than the exception. However, the systems analyst should watch for opportunities to replace probability functions with more accurate mathematical equations whenever possible.

Airline Ticket Counter

The application of stochastic modeling is explained best by example:

The market research department of an airline has determined that a major area of dissatisfaction among airline passengers is having to wait in line ten to twenty minutes for ticket purchases and luggage check-in. The management of the airline feels that the airline could gain a competitive advantage by significantly reducing this wait time and developing an advertising program to inform consumers of such a reduction.

To determine what measures are necessary to reduce the wait time, it is decided to develop a general-purpose model of the line-waiting process at the numerous ticket counters at various airports. Such a model can be used to simulate the wait time being

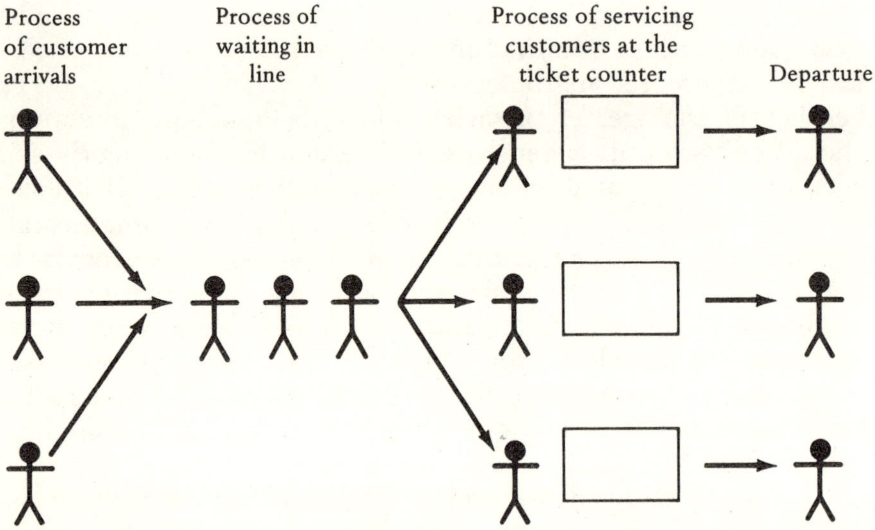

Figure 11-8. *Processes Associated with an Airline Ticket Counter*

encountered by passengers and then to experiment with ways to improve it. The key processes to be simulated are the arrival of customers, their time spent waiting in line, the time spent servicing customers, and the departure time of customers (see Figure 11-8).

Management is concerned with keeping line-waiting time to a minimum. This can be accomplished by having an adequate number of attendants and computer terminals available at the ticket counters. However, management also wants to avoid over-staffing ticket counters, resulting in idle employees and equipment, an unnecessary cost.

Probability Distributions and Random Numbers

The first step of stochastic modeling is to develop probability distributions for a process. These distributions can be based upon a frequency distribution from actual observations of the process, or they can be arrived at subjectively. For example, a frequency distribution of interarrival times (i.e., times since last arrival) of customers at an airline ticket counter is given in the first two columns of Table 11-2. The third and fourth columns of Table 11-2 illustrate the relative and cumulative frequencies of the observations. It is now possible to simulate arrivals of customers by the use of random numbers that are merely a quantitative reflection of chance. To be considered random, any number from a range of numbers must have an equal chance of occurring (e.g., as in rolling dice, spinning a roulette wheel, or drawing a number from a hat).

Table 11-2. *Observed Frequency Distribution of Interarrivals of Customers at an Airline Reservation Counter*

Interarrival Time (in minutes)	*Observed Frequency*	*Relative Frequency*	*Cumulative Frequency*	*Categories of Random Numbers*
1	15	15%	15%	1–15
2	20	20%	35%	16–35
3	30	30%	65%	36–65
4	20	20%	85%	66–85
5 and more	15	15%	100%	86–100
	100 *total*	100%		

Rather than draw numbers from a hat (as was done earlier in the chapter), the computer is used to generate random numbers. A computerized random number generator is simply a computer program that produces a sequence of numbers that appear to be random. Computer programs capable of generating a series of numbers having the statistical properties of random numbers are relatively easy to create.

A random number generator can quickly generate a stream of random numbers that can be translated into values for interarrival times by comparing the values to the cumulative probability distribution. To accomplish this, the possible range of the random numbers is set (from 1 to 100 in this example). The values of the random numbers actually generated are categorized into the ranges defined in the last column of Table 11-2. For example, a random number value of 6 falls between 1 and 15; therefore, it causes an interarrival time of 1 minute to be generated. Other random numbers produce interarrival times as follows:

Random Number	*Resulting Interarrival Time*
98	5
38	3
16	2
1	1
84	4

If the simulation model is functioning properly, a stream of 100 random numbers will result in a simulated frequency distribution that is very close to the one originally observed.

Once a model has been developed to simulate the arrivals of customers, the next step involves developing a model of the times required for ticket counter attendants to process customer tickets and luggage. This is accomplished using the same technique as used for developing a model for arrivals of customers. Once both models are operational, the arrival model can be used to feed simulated passenger arrivals to the ticket counter model. To the extent that the ticket counter model has adequate capacity to service all simulated passengers, no lines will develop. To the extent that the passenger arrival model gets ahead of the ticket counter model (i.e., generates simulated passengers faster than they can be serviced), lines will develop.

A computer program of this model can quickly simulate hours, days, weeks, or even years of passenger processing. Statistics can be readily gathered on the amount of time that customers spend in line and the amount of idle time that ticket counter attendants have.

Though the computer is just pretending to process passengers, the statistics about the simulation are representative of the statistics that would result from the real process. Management can use the model to conveniently experiment using different numbers of attendants and different arrival rates at the different airports. The results of their experimentation can be used to estimate the number of attendants and terminals needed at the airports to keep line-waiting times at an acceptable level (e.g., below three minutes).

Simulation Applications and Languages

The airline ticket counter simulation is concerned with the balancing between line waiting and idle capacity. A number of organizational processes fall into this type of simulation. Examples include processing of customer orders, checking out customers at supermarkets or retail stores, assembly line processes, unloading docks for freight trucks, and checking patients into hospitals.

Of course, stochastic simulation is applicable to any organizational process that can be described in terms of probability. For example, simulation can be used to model weather conditions, equipment failures, customer attitudes, and actions by competitors. Stochastic models can also be used in conjunction with deterministic models. For example, additional analyses may make it possible to develop mathematical equations to represent the process of servicing customers at an airline ticket counter. This deterministic model may be based upon how each customer pays (e.g., cash, credit card, or check), how many connecting flights must be scheduled, and how much luggage must be checked. The stochastic model of generating passenger arrivals could feed this deterministic model with simulated passenger arrivals, the same as it fed the original stochastic model of the ticket counter.

Initially, most simulation was programmed in FORTRAN, using random number subroutines to generate random numbers when needed. Though FORTRAN is still commonly used, a number of high-level simulation languages that minimize the difficulty of programming simulation models have been developed. These simulation languages are usually designed to simulate particular types of processes; they are of little value out of the areas for which they are designed. However, when a simulation language fits a situation, it is usually preferable to a language such as FORTRAN. For example, GPSS (General Purpose Systems Simulation) is a simulation language developed by IBM specifically for line-waiting problems. Other popular general-purpose simulation languages are SIMSCRIPT, DYNAMO, and GASP. It is considerably easier to simulate line-waiting problems and the like using simulation languages rather than FORTRAN.

Decision Design under Uncertainty

Decision making under uncertainty pertains to situations where alternative decisions can result in several identifiable outcomes, but their probabilities of occurrence are unknown. The potential contribution of a DSS is lower in this area of decision making.

In some cases, management may assign arbitrary probabilities to outcomes. For example, management may simply assign equal probabilities. When this is done, the problem can be treated as decision making under risk, and stochastic modeling can be used. However, caution is warranted in analyzing the results, because of possible inaccuracies in the probabilities assigned.

Without assigning arbitrary probabilities and using stochastic models, decision making under uncertainty is of necessity intuitive/heuristic. This type of decision making tends to be trial-and-error oriented. Therefore, the major contribution a DSS can make in this environment is to provide rapid feedback about outcomes after a decision is made, thus facilitating quick reactions to outcomes. This allows corrective decisions to be made if outcomes are undesirable. For example, there may be uncertainty about the demand for a new product. Therefore, its inventory level is determined on an intuitive basis. By monitoring the demand for the new product on a daily basis (i.e., with short-interval feedback), management can adjust inventory levels heuristically. This short-interval feedback allows a bad decision to be recognized and a corrective decision made.

DSS AND DECISION CHOICE

The choice phase of decision making involves selecting a particular course of action to resolve a problem or to capitalize on an opportunity. DSS support of decision choice results from the output of the models during decision design. The difficulty of the choice phase is, as might be expected, a function of the degree of certainty associated with outcomes.

Decision Choice under Certainty

If a decision is made under certainty, the process of choosing a course of action is to a great extent computational. Once the optimum solution has been determined, it is the rational choice.

Decision making under certainty is very similar to programmed decision making. In fact, once managers develop confidence in a model for a particular decision-making process, they often allow future decisions to be programmed, based on the solutions generated by the model.

Decision Choice under Risk

The output from a stochastic model varies each time the model is executed. Therefore, the usual approach is to execute the model repeatedly and develop probability distributions describing the outputs from the model. These statistics are used to develop probabilities of outcomes as they pertain to the real process. For example, management can determine that if there are four ticket counter attendants at a particular airport, there is a 90 percent probability that customers will never have to wait more than three minutes for service.

Decision making when the probabilities of outcomes are known is similar to decision making under certainty. However, instead of optimizing an outcome, the usual approach is to optimize the expected outcome. For example, assume a manager is faced with three choices, one offering a 10 percent probability of a gain of $100,000, the second offering a 20 percent probability of a gain of $45,000, and the third offering a 70 percent probability of a gain of $20,000. The expected values are computed as follows:

Alternative	Probability ×	Outcome =	Expected Value
1	.10	100,000	$10,000
2	.20	45,000	$ 9,000
3	.70	20,000	$14,000

Given these expected values, the third choice (offering an expected value of $14,000) would be selected.

Due to statistical properties, expected value as a selection criterion is most valid when the decision to be made is made a number of times. Since this is frequently not the case, there is evidence to suggest that behavioral factors influence decision making when a decision must be made only once. Often, depending on the optimism or pessimism a manager has, he or she will gamble that a lower-probability, but higher-valued outcome will occur. For example, an optimistic manager may select the first alternative in the preceding expected-value analysis.

Decision Choice under Uncertainty

Situations where management does not at least have subjective thoughts about the probabilities of different outcomes are fortunately rare. However, when they occur, decision choices must be made in an intuitive/heuristic manner.

To the extent that management is willing to assign arbitrary probabilities, stochastic models can be used to generate expected values. However, considering the arbitrary nature of these values, the process of choice should be combined with the intuitive feelings of management. When no probabilities are assigned, DSS technology is of no particular value for decision choice. Decision making must be made on a nonquantitative basis.

SUMMARY

Decision support systems (DSS) offer significant potential for assisting and improving management decision making. DSS support for the intelligence and design phases of decision making can lead to better choices with more predictable outcomes.

Decisions can be categorized as made under certainty, risk, or uncertainty. The category is determined by the intelligence phase of decision making. A DSS can provide intelligence through routine reporting, information retrieval capabilities, and statistical packages.

During decision design, computer-based numerical models can be used to develop and define decision alternatives. Deterministic models are used for decision design under certainty. Stochastic models are used for decision design under risk. DSS are limited in the contribution they can make to decision design under uncertainty. However, rapid feedback mechanisms can be used to inform the decision maker of the emerging outcome of his or her decision, thus allowing adjustments if a bad decision has been made.

The choice phase of decision making involves selecting a particular course of action. DSS support of decision choice results from the output of the models during decision design. When decision making is under certainty, the process of choosing a course of action is to a great extent based on the optimum solution as computed by a deterministic model. When decision making is under risk, the course of action selected is usually a function of the expected values of the possible outcomes. However, management may select a course of action with a lower or higher expected value, due to optimistic or pessimistic considerations. When decision making is under uncertainty, a DSS is of limited value.

EXERCISES

1. Discuss the characteristics of decision making and information used for decision making at operating-, middle-, and top-management levels.
2. Define decision making under certainty, risk, and uncertainty. Give examples of each.
3. Discuss behavioral characteristics of managers and their implications for the systems analyst.
4. Discuss how a DSS can assist management during the intelligence phase of decision making.
5. Differentiate between deterministic and stochastic models.
6. What is simulation? How can it assist in decision making?
7. Why are DSS limited in their support of decision making under conditions of uncertainty?

8. Discuss the use of rapid feedback to assist in decision making.

9. What are expected values? For what type of decision making (i.e., certainty, risk, or uncertainty) are expected values used?

SELECTED REFERENCES

Aiso, M. "Forcasting Techniques," *IBM Systems Journal* Vol. 12, No. 2 (June 1973), pp. 187–209.

Barrett, Michael. *Information Processing Types and Simulated Production Decision Making.* Working Paper 73-2. Minneapolis: The MIS Research Center, University of Minnesota.

Barton, Richard F. *A Primer on Simulation and Gaming.* Englewood Cliffs, N.J.: Prentice-Hall, 1970.

Bomball, Mark R.; James A. Hallam; Stephen F. Hallam; and Donald D. Scriven. "Understanding Simulation," *Data Management,* Feb 1975, pp. 30–32.

Boulden, J. B., and E. S. Buffa. "Corporate Models: On-Line Real-Time Systems," *Harvard Business Review,* July–Aug 1970, pp. 65–83.

Carlson, Eric D. "Proceedings of a Conference on Decision Support Systems," *Data Base* Vol. 8, No. 3 (Winter 1977).

Clark, Charles T., and Lawrence L. Schkade. *Statistical Analysis for Administrative Decisions.* 2d ed. Cincinnati: South-Western, 1974.

Davis, Gordon B. *Management Information Systems: Conceptual Foundations, Structure, and Development.* New York: McGraw-Hill, 1974.

Emshoff, J. R., and R. L. Sisson. *Design and Use of Computer Simulation Models.* New York: Macmillan, 1970.

Gordon, Geoffrey. *System Simulation.* Englewood Cliffs, N.J.: Prentice-Hall, 1969.

Gorry, Anthony G., and Michael Scott S. Morton. "A Framework for Management Information Systems," *Sloan Management Review* Vol. 13, No. 1 (Fall 1971), pp. 55–70.

Hallam, Stephen F.; Donald D. Scriven; Mark R. Bomball; and James A. Hallen. "Basic Steps in Developing Simulation Models," *Data Management,* Apr 1975, pp. 26–29.

Hammond, John S., III. "Do's & Don'ts of Computer Models for Planning," *Harvard Business Review,* Mar–Apr 1974, pp. 110–123.

Infosystems Staff. "Management Gets the Picture," *Infosystems,* Apr 1977, pp. 37–42.

Jones, C. H. "At Last: Real Computer Power for Decision Makers," *Harvard Business Review,* Sept–Oct 1970, pp. 75–89.

Keen, Peter G. W., and Michael Scott S. Morton. *Decision Support Systems: An Organizational Perspective.* Reading, Mass.: Addison-Wesley, 1978.

Kim, Chaiho. *Quantitative Analysis for Managerial Decisions.* Reading, Mass.: Addison-Wesley, 1976.

Kingston, P. L. "Concepts of Financial Models," *IBM Systems Journal* Vol. 12, No. 2 (June 1973), pp. 113–125.

12 Future Considerations of Systems Analysis

In this last chapter of the book, the future of systems analysis is discussed. Future organizations and computer technology are discussed in the first two sections of the chapter. The third section discusses the implications of future organizations and computer technology on systems analysis.

FUTURE ORGANIZATIONS

It is difficult to attempt to forecast the future development of organizations. A multiplicity of unpredictable and unforeseen forces will shape future organizations. For example, few organizational forecasts made in the 1950s predicted the "counterculture" movements of the 1960s and 1970s—the unanticipated rejection of many basic values of our society. People, particularly young people, began to question previously accepted values such as the Protestant work ethic, the pursuit of materialistic betterment, esteem for scientific and technological achievements, and the belief that economic growth is always a good thing. These counterculture movements have affected the way in which organizational participants view work and organizations. Organizations, both government and business, have been increasingly subject to criticism. They have often had to alter their approaches to the production of goods and/or services in terms of both what they produce and how they produce it.

Because of the unpredictability of the future, it is presumptuous to attempt to describe future organizations without emphasizing the possibility of error in the forecasting. However, it is reasonable and worthwhile to make a few generalized forecasts about future organizations. Such forecasts are useful in preparing for the future. At least some, if not many, of the forecasts are likely to be common.

There are two basic factors to consider in discussing future organizations. One is the environment in which organizations operate; the second is the actual operation of organizations. These factors are discussed below.

Environment of Future Organizations

Organizations do not operate in a vacuum. Rather, they are dependent upon, and must interact with, their environment (e.g., customers, competitors, governments, and natural resources). Accordingly, organizations must be cognizant of changes in their environment and make appropriate adjustments in order to survive.

Market Conditions

Traditionally, the primary environmental concern of organizations has been to recognize and respond to changing market conditions. For example, an organization producing buggy whips in the early 1900s had to recognize that the proliferation of automobiles would displace the need for buggy whips. Consequently, the organization had to diversify its production efforts into other goods and/or services, or face extinction.

Increasingly, the time in which an organization must recognize and adjust to changing market conditions is narrowing. Consider organizations in the fashion, ski equipment, computer, and electronic calculator industries. Popular products can quickly lose their market appeal as a result of new products released by competitors. For example, the NCR company, the leading producer of cash registers, temporarily lost its domination of the retail business machine market to the Singer company due to the advent of electronic retail terminals. In the intense competition that ensued, NCR regained its position in the market and Singer withdrew from the market completely. Had NCR not recognized and responded to the changes in market conditions in sufficient time, the organization may not have regained its market position; it may even have gone out of business.

It is reasonable to assume that future organizations will have to cope with increasingly unpredictable and less forgiving market conditions. Therefore, organizations not only are going to have to more quickly recognize changes in the marketplace, but also are going to have to be more flexible to respond to these changes.

Society and Natural Resources

There are other environmental concerns that present challenges to organizations in the future. These challenges can be broadly categorized as pertaining to society and natural resources. From a social perspective, organizations must respond increasingly to expanding social pressures. If an organization fails to respond, these social pressures may manifest themselves in the form of rejection of the organization's products and/or services, government legislation, and even sabotage. Organizations of the future will be expected to assume (or will be legislated into assuming) greater social responsibility for such things as pollution, consumer safety, equality of employment, and retraining of workers displaced by new technologies. To keep social pressures from becoming unreasonable, organizations will have to influence public opinion and legislation increasingly in order to protect their interests.

The shortage of natural resources has become, and will continue to be, a problem. Early economic theory assumed an unlimited supply of natural resources. This assumption is obviously not true. Organizations are going to have to cope with these shortages and participate in, or at least support, efforts to discover and operationalize new forms of energy and materials.

Operation of Future Organizations

When organizations are viewed in terms of their internal operations, certain forecasts can be made.

Size and Complexity

Organizations are likely to continue to increase in size and complexity. Organizations are expanding both the market for their existing products and their product offerings. One of the most visible areas in which organizations are expanding existing product activities is in the international market. For example, a large portion of IBM's computer sales are transacted outside of the United States. The multinational organization has become a common phenomenon that is likely to continue in the future.

Organizations are expanding the products and/or services they offer by diversifying their efforts into new and varied fields. Consider the diverse products and/or services offered by such organizations as General Electric, DuPont, General Motors, the Federal Government, and higher education. Universities, for example, have significantly expanded their academic programs into areas not even recognized a few years ago.

Technology

The technology used by organizations to produce goods and/or services has become, and will continue to be, more complex and less stable. Technology is used in this context to refer not only to the equipment used in organizations, but also to the skills of organizational participants (engineers, computer technicians, accountants, lawyers, etc.).

The field of computing and information systems provides a vivid example of the increasing complexity and instability of technology. Computer hardware has evidenced life cycles of from five to ten years, and programming skills have been affected in a parallel fashion. Programming skills have been upgraded from board wiring in the 1950s, to batch programming in the early 1960s, to online programming in the late 1960s, to database programming in the 1970s.

To cope with the dynamics of technology in all dimensions of the organization, it will be necessary to forecast future technologies and their impacts on the organization, just as it is necessary to forecast market conditions. If an organization fails to consider future technology in the purchasing of equipment and the selection and training of personnel, its survival may be seriously threatened.

FUTURE COMPUTER TECHNOLOGY

Predicting future computer technology is as risky as predicting future organizations. To illustrate, in 1950, computer and industry experts predicted the total number of computers that would be manufactured would be 10. This forecast was obviously in error. There are several hundred thousand computers installed today. However, considering the size, cost, and difficulty of using early computers, the forecast was not unrealistic. What the forecasters were unaware of was the great strides forthcoming in the technology used to manufacture computers. Computers today can perform calculations, store data, and be programmed for a minute fraction of the cost and difficulty of equivalent calculations, data storage, and programming of computers manufactured in the 1950s.

Improvements in the cost effectiveness of both hardware and software continue to expand the frontiers of possible computer applications. Some hardware and software developments likely to occur in the future are discussed below.

Hardware

Computer Processors

Computer processors of the future will be constructed of smaller, faster circuits. During the late 1980s, computer circuits with performances at least ten times faster and one-tenth as expensive as those in use in the late 1970s will be available.

Lower-cost, higher-performance processing units will result that not only will proliferate the use of microcomputers, minicomputers, and large computers as stand-alone devices, but also increase the tendency to combine several processors into multicomputers. In the multicomputer environment, multiple processors (of varying sizes) perform different tasks traditionally performed by one processor. For example, processors are assigned to handle different input and output functions. This multicomputer architecture significantly increases the overall performance capabilities of a single computer system. It also allows a system to be totally backed up by a second set of processors that can be activated if one or more of the primary processors malfunctions.

File Storage

The ability to store massive databases will be significantly improved in future years due to improvements in magnetic disk technology. It appears reasonable to expect a twenty-fold increase in the amount of data that can be stored on magnetic disk surfaces by the late 1980s.

Besides improvements in disk technology, considerable research is being conducted into other file storage technologies—magnetic bubbles, laser-holographic devices, charge-coupled devices, cryogenic devices, and others. This research will likely result in new types of file-storage devices that will augment, and possibly replace, magnetic disk devices.

Input/Output

Little dramatic improvement is anticipated in the basic technologies of input/output devices. Electromechanically oriented devices such as card readers and line printers are relatively mature as far as their technologies go. However, higher-performance, nonimpact printers show real promise. Devices ranging from low-cost thermal printers to higher-cost electrostatic and ink-jet printers should eventually displace impact printers. The electrostatic and ink-jet technologies offer additional advantages and greater versatility—for example, multiple colors, multiple fonts, and graphic capabilities.

Computer-output microfilm (COM) will be used increasingly to alleviate problems associated with distributing and storing computer printouts. Also, word-processing equipment will continue to increase in use as information systems expand from traditional data processing activities into office environments. A novel input/output technology that will increase in use is voice recognition.

The most significant input/output technology of the future will be remote terminals. In this area, it appears that the CRT (cathode-ray tube) will dominate for some time because of its versatility and low cost. However, hardcopy typewriterlike terminals will also increase in use. One of the more interesting trends in terminals is their increasing intelligence. The use of microcomputers in terminals will grow as terminals perform more processing and editing functions. More application-oriented terminals (e.g., banking and retailing devices) can be expected as greater versatility in terminals evolves.

Software

Software is the area where major computer technology accomplishments are likely to occur. Software capabilities have always lagged behind hardware capabilities. Hardware technology appears to be stabilizing; it is reasonable to expect an increasing emphasis on software technology.

Major software advancements can be anticipated in the areas of database management, data communications, and programming languages. These topics are discussed below.

Database Management

The various "housekeeping" functions of managing data, records, and files will be simplified increasingly by advancements in database technology. Such functions as disk track and cylinder management, secondary indexing, space allocation, address randomization, and staging of data volumes in the hierarchy of devices will eventually become functions of database management systems. These functions will be as transparent to application programmers as the various functions performed by operating systems are today. Programmers will become more productive as they are relieved from these time-consuming tasks.

Data Communications

Data communications management functions such as adapting for varied line, terminal, and message characteristics will be increasingly accommodated by front-end controllers and systems management software. This will simplify the interfacing required for online networks in which a wide variety of computers and terminals are linked together. The automatic handling of data communications protocols will save considerable programmer time and simplify systems design.

Easier Programming

Support of such concepts as top-down (or outside-in) design, structured programming, and semi-automatic documentation will become common. There will also be increasing support for conversational debugging of computer programs.

These and other capabilities should alleviate increasing systems development costs. Improvements of up to 30 percent in programming productivity should be possible with these programming aids.

Improvements in general-purpose query languages will expand the use of computers by persons other than computer technicians. Managers will be able to obtain the information they need by directly interacting with computers.

IMPLICATIONS FOR SYSTEMS ANALYSIS

The future developments of organizations and computer technology will impact what systems analysts do and how they do it. Several major implications are discussed below.

Organizational Implications

Environment

Future organizations are going to be faced with a more turbulent environment in terms of market conditions, society, and natural resources. To remain viable, organizations are going to require more information about their environment in order to make sound decisions. Computer-based information systems will continue to expand beyond the scope of processing information based on activities internal to the organization, to processing information based on activities external to the organization.

Since it is not usually practical to collect and process data about the entire environment of an organization, statistics will be increasingly used to infer changes in the environment. For example, there will be increasing emphasis on marketing research and forecasting.

A single organization can incur considerable expense in obtaining information external to the organization. Therefore, the collection and maintenance of a multiplicity of databases about such things as market conditions and labor availability in particular industries is likely. For example, an independent company may develop a database of consumers and their preferences for the motel and hotel industry. Various motel and hotel chains will then be able to purchase information extracted from the database much more economically than they can collect and process it themselves.

One of the more critical environmental concerns for organizations is the increasing interest in, and pressure for, privacy legislation. Legislation to protect individuals from the abusive invasion of privacy possible through computer technology is continually being proposed and debated. Though individual privacy needs to be protected, the means by which it is protected can have serious implications for organizational databases. Organizations may be required to delete certain data from their databases, restrict the exchange of data among other organizations, and provide tighter security. They may also be subjected to more stringent audits. From a technical perspective, such legislation can severely impact both the effectiveness and the efficiency of computer-based information systems.

Internal Operations

The internal operations of future organizations can be characterized as expanding. This expansion is resulting in increasing organizational size and complexity. The technology used to produce goods and/or services is becoming less stable and less predictable. It is going to be both more difficult and more important to integrate the information about the more diversified and more volatile organizational activities in order to maintain management control and coordination.

Computer Technology Implications

Faced with the increasing need for information by organizations, it is encouraging to note that future computer technology is very promising. The projected reductions in the cost of hardware will expand the number and scope of viable areas in which computer technology can be applied. The declining cost and increasing performance of computer hardware will make it possible for the hardware to cost effectively absorb the overhead required to support the projected software enhancements (e.g., database management systems) without seriously degrading computer performance.

Clearly, more of the traditional technical details of computer technology will be handled by hardware and software supplied by vendors. Consequently, systems analysts will be able to concentrate more on the information requirements that must be satisfied by the information systems and less on computer technology. Systems analysts and computer programmers will have to accept the obsolescence of many of their technologically elegant skills.

The systems analysts of the future will become more involved in understanding organizations and organizational processes in order to assist organizations in determining how to logically integrate information and how to develop decision-support systems. Systems analysts will be less involved in the intricacies of the computer technology needed to support information systems.

SELECTED REFERENCES

DeCarlo, Charles R. "Changes in Management Environment and Their Effect upon Values," in *The Impact of Computers on Management.* ed. Charles A. Meyers. Cambridge, Mass.: The M.I.T. Press, 1967.

Johnson, Richard A.; Fremont E. Kast; and James E. Rosenzweig. *The Theory and Management of Systems.* New York: McGraw-Hill, 1973.

Hunt, J. G., and P. F. Newell. "Management in the 1980's Revisited," *Personnel Journal* Vol. 50, No. 1 (Jan 1971), pp. 35–43, 71.

Kast, Fremont E., and James E. Rosenzweig. *Organization and Management: A Systems Approach.* New York: McGraw-Hill, 1974.

Kriebel, Charles H. "The Future MIS," *Business Automation* Vol. 19, No. 6 (June 1972), pp. 18–19.

Leavitt, Harold J., and Thomas L. Whisler. "Management in the 1980's," *Harvard Business Review,* Nov–Dec 1958, pp. 41–48.

Nanus, Bust. "Managing the Fifth Information Revolution," *Business Horizons,* Apr 1972, pp. 5–6.

Roman, Daniel O. "Technology Forecasting in the Decision Process," *Academy of Management Journal,* June 1970, pp. 127–138.

Sayles, Leonard R., and Margaret K. Chandler. *Managing Large Systems: Organizations for the Future.* New York: Harper & Row, 1971.

Withington, Frederic G. "Five Generations of Computers," *Harvard Business Review,* July–Aug 1974, pp. 99–108.

Withington, Frederic G. "Future Computer Technology," *Data Base* Vol. 7, No. 4 (Spring 1976), pp. 7–14.

Index